Spiros Sirmakessis (Ed.)

Adaptive and Personalized Semantic Web

T0141984

Studies in Computational Intelligence, Volume 14

Editor-in-chief
Prof. Janusz Kacprzyk
Systems Research Institute
Polish Academy of Sciences
ul. Newelska 6
01-447 Warsaw
Poland
E-mail: kacprzyk@ibspan.waw.pl

Spiros Sirmakessis
(Ed.)

Adaptive and Personalized Semantic Web

 Springer

Dr. Spiros Sirmakessis

Research Academic
Computer Technology Institute
Riga Fereou 61
26221 Patras
Greece
E-mail: syrma@cti.gr

ISSN print edition: 1860-949X
ISSN electronic edition: 1860-9503

ISBN:13 978-3-642-06785-3
e-ISBN:13 978-3-540-33279-4

Springer is a part of Springer Science+Business Media
springer.com
© Springer-Verlag Berlin Heidelberg 2006
Softcover reprint of the hardcover 1st edition 2006

Foreword

Web Personalization can be defined as any set of actions that can tailor the Web experience to a particular user or set of users. To achieve effective personalization, organizations must rely on all available data, including the usage and click-stream data (reflecting user behaviour), the site content, the site structure, domain knowledge, as well as user demographics and profiles. In addition, efficient and intelligent techniques are needed to mine this data for actionable knowledge, and to effectively use the discovered knowledge to enhance the users' Web experience. These techniques must address important challenges emanating from the size and the heterogeneous nature of the data itself, as well as the dynamic nature of user interactions with the Web. These challenges include the scalability of the personalization solutions, data integration, and successful integration of techniques from machine learning, information retrieval and filtering, databases, agent architectures, knowledge representation, data mining, text mining, statistics, user modelling and human-computer interaction. The Semantic Web adds one more dimension to this. The workshop will focus on the semantic web approach to personalization and adaptation.

The Web has been formed to be an integral part of numerous applications in which a user interacts with a service provider, product sellers, governmental organisations, friends and colleagues. Content and services are available at different sources and places. Hence, Web applications need to combine all available knowledge in order to form personalized, user-friendly, and business-optimal services.

The aim of the International Workshop on Adaptive and Personalized Semantic Web that was held in the Sixteenth ACM Conference on Hypertext and Hypermedia (September 6-9, 2005, Salzburg, Austria) was to bring together researchers and practitioners in the fields of web engineering, adaptive hypermedia, semantic web technologies, knowledge management, information retrieval, user modelling, and other related disciplines which provide enabling technologies for personalization and adaptation on the World Wide Web.

Topics of the Workshop include but are not limited to:

- design, methodologies and architectures of adaptable and adaptive Web information systems
- user interface design for adaptive applications on the Web
- semantic web techniques for adaptation
- authoring of adaptive hypermedia for the Web
- distributed user modelling and adaptation
- semantic web mining
- personalized taxonomies or ontologies
- hybrid recommendation systems
- model integration for personalization and recommendation systems
- web usage, content, and structure mining
- automated techniques for generation and updating of user profiles
- machine learning techniques for information extraction and integration
- applications of relational data mining in personalization
- adaptive personalized web applications

This workshop could not have been held without the outstanding efforts of Marios Katsis at the workshop support. I would like to thank the programme committee members for their efforts and support. Finally, recognition and acknowledgement is due to all members of the Internet and Multimedia Research Unit at Research Academic Computer Technology Institute.

Dr Spiros Sirmakessis
Assistant Professor
R.A. Computer Technology Institute
syrma@cti.gr

Programme Committee and Reviewers

Bamshad Mobasher, School of Computer Science, Telecommunication, and Information Systems, DePaul University, USA.

Olfa Nasraoui, Dept of Computer Engineering & Computer Science, the University of Louisville, USA.

Lars Schmidt-Thieme, Computer-based New Media, Institute for Computer Science, University of Freiburg, Germany

Martin Rajman, Center for Global Computing, EPFL, Swiss Federal Institute of Technology, Switzerland.

Ronen Feldman, Department of Computer Science, Bar-Ilan University, Israel.

Spiros Sirmakessis, Computer Technology Institute and Technological Educational Institution of Messolongi, Greece.

Peter Haase, Institute AIFB, University of Karlsruhe, Germany.

Michalis Vazirgiannis, Department of Informatics, Athens University of Economics & Business, Greece.

Ioannis Hatzilygeroudis, Computer Engineering and Infomatics Department, University of Patras and Computer Technology Institute, Greece.

Steven Willmott, Languages and Systems Department, Universitat Politecnica de Catalunya, Spain.

Michalis Xenos, Hellenic Open University, Greece.

Miltiadis Lytras, Computer Engineering and Infomatics Department, University of Patras and Computer Technology Institute, Greece.

Contents

htmlButler – Wrapper Usability Enhancement
through Ontology Sharing and Large Scale Cooperation

A Methodology for Conducting Knowledge Discovery
on the Semantic Web

An Algorithmic Framework
for Adaptive Web Content

Christos Makris, Yannis Panagis, Evangelos Sakkopoulos and Athanasios
Tsakalidis

Research Academic Computer Technology Institute, RU5, Patras, Greece
Dept. of Computer Engineering and Informatics, University of Patras, Greece
{makris,sakkopul,panagis,tsak}@ceid.upatras.gr

Abstract. In this work a twofold algorithmic framework for the adaptation of web
content to the users' choices is presented. The main merits discussed are a) an optimal offline site adaptation – reorganization approach, which is based on a set of different popularity metrics and, additionally, b) an online personalization mechanism
to emerge the most "hot" (popular and recent) site subgraphs in a recommendation
list adaptive to the users' individual preferences.

1 Introduction

User driven access to information and services has become more complicated,
and can sometimes be tedious for users with different goals, interests, levels of
expertise, abilities and preferences. The Boston Consulting Group announced
that a full of 28% of online purchasing transactions failed and 1/3 of them
stopped shopping on-line due to usability difficulties [2]. This problem is crucial in on-line sales systems with thousands of products of different kinds
and/or categories. It is obvious that typical search methods are becoming less
favorable as information increases resulting in money losses.

In user-centered applications, two parameters affect usability:

- Orientation and navigation strategy. Users are frequently uncertain as how
  to reach their goals. Since users have different states of knowledge and
  experience, information presentation may be too redundant for some of
  them and too detailed for others.
- Quality of search results. Users cannot locate efficiently the information
  they need (results must be relevant and come quickly).

Moreover, with the unprecedented growth of the Internet usage, websites
are being developed in an uncontrollable, ad-hoc manner, a fact frequently
reflected to unpredictable visit patterns. Thus, a critical task for a website

C. Makris et al.: *An Algorithmic Framework for Adaptive Web Content*, Studies in Computational Intelligence (SCI) **14**, 1–10 (2006)
www.springerlink.com

maintainer is to use enumerable metrics in order to identify substructures of the site that are objectively popular.

Web Usage Mining has emerged as a method to assist such a task. The fundamental basis for all mining operations entails processing web server access logfiles. In its most simplified approach, usage mining entails registering absolute page visits or identifying popular paths of information inside a website, by the means of logfile analysis software solutions such as Webtrends (http://www.webtrends.com), and Analog (http://www.analog.cx). When the goal is to detect popular structural website elements, more elaborate techniques have been devised. Some representative work is presented hereafter.

This work contributes two main approaches: it presents an optimal offline site adaptation – reorganization approach based on a set of different popularity metrics and it presents an online personalization mechanism to display the most "hot" – popular and recent – site subgraphs in a recommendation list adaptive to the users' individual preferences. Both approaches build on well-known results in data structures in the areas of optimal trees and adaptive data structures.

The rest of the paper is organized as follows. Section 2 present background and related work. Section 3 presents popularity metrics that can be taken into account after analyzing user behaviour. Metrics are both localized, i.e. for certain pages and more globalized. Section 4 presents two approaches to reorganize website structure after having computed the appropriate metrics. We conclude with future directions.

2 Background and Related Work

To receive web usage feedback, web sites have been accompanied with logging mechanisms that have been evolving over time. However, the consequences of ad hoc implementation are depicted on the logged navigation trails, where mining for useful information in the logs has become a travel through an information labyrinth.

A shift from standard HTML based applications toward server side programmed web applications is noted several years now, especially with the advent of technologies such as Java servlets, PHP and lately with Microsoft.NET. Among other features, new techniques allow URL re-writing to provide additional information on the HTTP requests, HTML server-side pre-rendering or pre-compilation to facilitate quicker download, client-side code injection to enable transparent capturing of additional user actions and custom logging databases that keep details regarding content delivery to web and proxy servers.

Significant work on converting server logfiles to valuable sources of access patterns has been conducted by Cooley [6]. Apart from analysing logfiles, it is important to use analysis as input and determine which changes, if any, to bring to the website structure. Chen et al. [3] describe efficient algorithms to infer access patterns corresponding to frequently traversed, website paths. Apart from analysing logfiles, it is important to use analysis as input and determine which changes, if any, to bring to the website structure. Srikant and Yang [13] infer path traversal patterns and use them to indicate structural changes that maximize (or minimize) certain site-dependent criteria. Finally, in [4, 5] the authors define techniques to assess the actual value of webpages and experiment on techniques and mechanisms to reorganize websites.

3 An Overview of Metrics for Webpages and Site Subgraphs

3.1 Access Smells: Absolute, Relative, Spatial and Routed Kinds

Several different metrics have been proposed to calculate the access frequencies from log file processing (see [7] for an early analysis on web logs). In this section, we present *access smells*, which are different kinds of metrics to estimate a web page's popularity. We present the approaches of [4] and [8].

As *absolute* kind of calculation, we refer to the *Absolute Accesses* (AA_i) to a specific page i of a site. The *relative* kind has been initially outlined in [8]. It is defined as:

$$RA_i = a_i * AA_i$$

That is, the RA_i of page i is a result of the multiplication of AA_i by a coefficient a_i. The purpose of a_i is to skew AA_i in a way that better indicates a page's actual importance. Hence, a_i incorporates topological information, namely page depth within site d_i, the number of pages at the same depth n_i and the number of pages within site pointing to it r_i. Thus $a_i = d_i + n_i/r_i$. According to [8], the number of hits a page receives, as those are calculated from log file processing, is not a reliable metric to estimate the page's popularity. Thus this refined metric is proposed, which takes into account structural information. Based on this new notion of popularity, reorganization of certain pages is proposed.

In [4] two more kinds have been introduced towards a similar direction. Acknowledging the importance of reorganization proposals, the aim was to further facilitate the idea of reorganization by introducing two new metrics. The first one takes into account both structural information and the differentiation between users coming from within the website and users coming from other websites, while the second uses a probability model to compute a suitable refining factor. Key feature of the new metrics is the higher fidelity on the proposed reorganization proposals. In this direction the authors decompose AA_i into two components, $AA'_{i,in}$ and $AA_{i,out}$ to account for absolute accesses from inside the site and from the outside world Thereby,

$$RA_i = a_{i,in} \cdot AA'_{i,in} + a_{i,out} \cdot AA_{i,out}$$

We call these metrics the *spatial* and *routed* kind of page popularity, respectively. *Spatial* kind is based on the topological characteristics of a web site. In particular, these characteristics are implied by the fact that a web page may be accessed using four different ways. Firstly it gets accesses originating from the site (neighboring pages), secondly directly via bookmarks stored at a client browser (indicating that a client prefers this site for his/her reasons), thirdly by incoming links from the outside world and finally by typing directly its URL. By considering accesses from inside the site we obtain $AA'_{i,in}$. Accounting the last three of the above access methods and with the proper normalization (see [4]) we obtain $AA_{i,out}$.

In the *routed* kind of calculation as presented in [4], the idea was to increase page relative weight, inversely proportional to its access probability. Considering a site structure as a *directed acyclic graph* (DAG) G, with s nodes and v_i denoting page i. Suppose that a user starts at the root page v_r, looking for an arbitrary site page v_t. At each node v_i he makes two kinds of decisions: either he stops browsing or he follows one of the $out(v_i)$ links to pages on the same site. If we consider each decision equiprobable, the probability p_i, of each decision is $p_i = (out(v_i) + 1)^{-1}$.

Consider a path $W_j = <v_r, v_1, \ldots, v_t>$, from v_r to v_t. Counting the routing probabilities at each step, the probability of ending up to v_t via W_j, is simply:

$$P_{t,j} = \prod_{\forall i, v_i \in W} p_i$$

There may be more than one paths leading to t, namely $W_1, W_2, \ldots, W_k$. The overall probability of discovering t, D_t is:

$$D_t = \sum_{i=1}^{k} P_{i,k}$$

For the example in Fig. 1,

$$D_2 = \frac{1}{3} \cdot \frac{1}{5} \cdot \frac{1}{3} + \frac{1}{3} \cdot \frac{1}{5} \cdot \frac{1}{2} + \frac{1}{3} \cdot \frac{1}{2} \cdot \frac{1}{3} \cdot \frac{1}{2}$$

Considering page i as target, the higher D_i is the smaller $a_{i,in}$ shall be, thus we choose $a_{i,in}$ to be, $a_{i,in} = 1 - D_i$. We also let $a_{i,out}$ to be one.

Thus we define RA_i as:

$$RA_i = (1 - D_i) \cdot AA'_{i,in} + AA_{i,out}$$

with $AA'_{i,in}$ and $AA_{i,out}$ defined as previously.

3.2 Recording User Visits and Hot Subgraphs

The mechanism of [1] scans the website graph in order to keep only the non-intersecting subpaths. Suppose that the site is modelled as a graph $G(V,E)$ kept in an adjacency matrix representation, with matrix A.

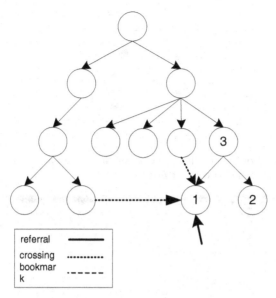

Fig. 1. The DAG structure of a simple site

After the completion of the identification of "Maximum Forward Paths" [1], the website access sequences (paths) are kept. Each path sequence P_i, of access frequency count (P_i), occurs with count $(P_i) \geq \min\{\text{support}\}$. In order to keep only these paths we set to zero the corresponding cells in A, thereby pruning the remaining paths. Suppose that there are p frequent paths. An array $C[1\ldots p]$ is kept, such that $C[i] = \text{count } (P_i)$ and p is the maximum number of paths.

A breadth-first search is initialised. Suppose we visit node v. If there is only one outgoing edge from v in the graph remaining after the initial pruning, then there must be a single non-zero element in the v-th row of A. If this is the case, we delete node v by zeroing out the non-zero entry. There is a small subtlety here; v might be part of a larger path. Therefore, we delete v only if there is no entry but zeros in the v-th column of A (an indication that no edge ends at v). When deleting node v we add the path frequencies of the paths passing through v, to their subpaths. For example suppose that we had a path $P_i= abfh$ and paths $P_j= bfh$, $P_k= bh$. After elimination of a we perform $C[j] = C[j] + C[i]$ and $C[k] = C[k] + C[i]$. The procedure stops when there are no nodes left to be eliminated. The remaining non-zero elements correspond to a subgraph of G with high popularity. For a non-zero element i, $C[i]$ keeps a popularity metric for path i. Actually what remains after this procedure is popular website paths that have common origins, see Fig. 2.

The above algorithm can gracefully adjust to multiple levels of granularity. After setting the threshold at the first path-elimination step, then after each vertex-elimination, one can define a higher threshold to correspond to more

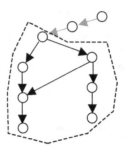

Fig. 2. Part of the output of the above algorithm. Grey arcs are discarded while the important part is enclosed in *dashed line*

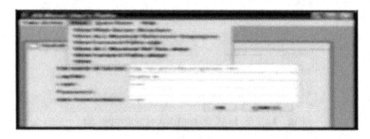

Fig. 3. Visualization Tool Capabilities (from [1])

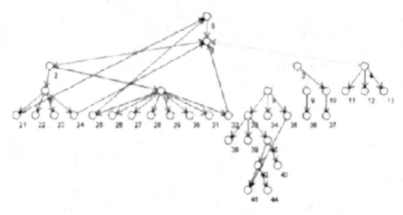

Fig. 4. Maximal Path Visualized Paradigm (from [1])

important paths, and so on. Using the visualization tool of Fig. 3 the three most popular paths are coloured in Fig. 4.

The process is repeated for any registered user in order to record into a user based profile the necessary information of his/her previous activity. In order to produce an adaptive recommendation list of the "hot" subgraphs that a user regularly visits the following adaptive data structure is adopted to provide automatic online site re-organization facilities.

4 Algorithms for Organizing Web Content According to Mined Access Patterns

In this section we present two approaches to be used as a second step after the initial computation of page importance. The results of the first stage are recorded into an adaptive user profile (outlined in [9]) to achieve personalization effects for registered users that visit web sites of interest. The first of them, *the offline*, uses computed importance weights, to optimally reorganize the structure of a website so that it minimizes the navigation entropy. The second one, *the online*, adapts the page presentation after each visit to a certain page.

4.1 Some Preprocessing Steps

In order for the following techniques to operate correctly we require the following preprocessing steps to take place.

Consider a metric M over some object i, and let m_i denote the corresponding value. We require M to define a probability distribution. Therefore, we require an initial normalization step to take place, i.e. we set $m_i = m_i \big/ \sum_{\forall j} m_j$.

Furthermore, the data structures used, require their items to be items of an ordered set. In order to satisfy this requirement, we traverse the site structure (in BFS or DFS) and assign each page a unique number. In the case of graph components of Sect. 3.2, we can use their unique path identifier instead.

Therefore, from now and on we will assume that we work on a set of website elements, single web pages or website components, each of which has been assigned a unique number and a normalized popularity metric.

4.2 Offline Optimal Reorganization Algorithm

After logfile processing, calculating one of the metrics of Sect. 3 and performing the process described at the previous subsection, we have arrived at the following situation: we have a set W of web elements and a probability distribution on W. A probability $p(w_i)$, $w_i \in W$ indicates the popularity of w_i. We would like to organize W in order to minimize the access entropy on its elements, i.e. our goal is to minimize

$$E = \sum_{\forall i} p(w_i) \cdot l_i$$

where by l_i we denote, the path length leading to element w_i. This problem is easily reduced to the problem of finding an optimal binary search tree. The most suitable version of this problem in the context of website reorganization, is the construction of a node-oriented search tree. This implies that we allow every node to contain useful information and not only the tree leaves as is the case in leaf-oriented trees. Hence, the equivalent problem is the following:

Given a set of n values $W = w_1 < w_2 < \ldots < w_n$ and an access distribution $D = \{p(w_1), p(w_2), \ldots, p(w_n)\}$ on its elements find the node-oriented search tree that minimizes E.

This problem has been addressed by Knuth [10]. The problem is solved by the means of a modified dynamic programming technique. Straightforward dynamic programming results in a solution with $O(n^3)$ time. However, after constraining an internal condition at the dynamic programming loop the time complexity drops to $O(n^2)$ in $O(n^2)$ space. For more details also refer to Mehlhorn [11], Sect. 3.4.

In order to automatise link creation after construction of an optimal tree structure, the framework introduced in [5] can be of great assistance.

We can extend the previous solution by embedding splay tree ([12]) heuristics while maintaining and reorganizing the structure of the Web Site. A splay tree is a binary search tree in which all operations are performed by means of a primitive, the so called *splaying* operation. A splay operation on a node x consists of a sequence of rotations from the node x to the root of the tree with each rotation rotating some higher edges before some lower edged and the whole sequence of a rotations corresponding to a splay leads to decreasing the depth of the nodes in a path by about a half. A crucial property that splay trees have is that they are statically optimal.

In order to apply splay trees in our setting we have firstly to transform the tree structure of the Web Site into a binary tree which is easy to do by replacing each node with degree k by a $\log k$ height binary tree. Hence each node of this tree has a weight corresponding to each importance metrics and the weight of each node is updated at a rate that is settled by the Web Site Administrator. Whenever the importance metric of a node is updated then the whole path from the node to the root is updated and so changes to the Web Site design can take place in the path from the leaf to the root. Details of how to implement this strategy are omitted from this short version of the paper.

4.3 Online Personalization Using Adaptive Data Structures

The previous approach was static in the sense that access results are gathered after a period of observation, access metrics are computed and then restructuring is performed. In many cases though, it is preferable to adapt content online and, e.g. give user a link table containing the most frequently and most recently accessed web site parts. A simple and elegant strategy to achieve this goal, without even the need to know the specific popularity of certain web elements, is to use an adaptive data structure. In the following we constrain for the sake of clarity our discussion to web pages.

The data structure that can be used is the *adaptive list*. The adaptive list is a doubly-connected list of unordered items. Each time an item is accessed, it is brought to the front (left end) of the list. This adaptation rule is called *Move-to-Front*. An example is shown in Fig. 5.

2, 7, 5, 1, 4, 3, 6 --- Access (3) ---> 3, 2, 7, 5, 1, 4, 6 --- Access (7) ---> 7, 3, 2, 5, 1, 4, 6

Fig. 5. The Move-To-Front rule

It is proved (see e.g. [11]) that Move-to-Front is at least 2-competitive, i.e. the total running time for a sequence of element accesses is at most twice slower than the optimum adaptation strategy. Note that the optimum strategy has full knowledge of the element distribution, whereas the adaptive list achieves its competitiveness without any knowledge of the distribution.

The adaptive list and the Move-To-Front rule can be the structure of choice whenever we want to keep a recommendation list of the most recently visited pages. These lists are kept for each web page and for each user in a personalized web environment and present users with possible navigation choices. In a possible implementation we can present users the leftmost k elements of the list, where k is a predefined constant. This amounts to presenting user with the k pages that she is most likely to visit in the future.

5 Conclusions and Future Steps

In this paper we have proposed two different approaches to adapting online content. Both our approaches are based on established results from the data structures' area and have aimed to provide a new viewpoint to a classical concern in online content. Those approaches are both elegant and easy to implement. Future steps include the description of a framework that it would evaluate the combination of reorganization metrics with different sets of redesign proposals. We also consider as open issue the definition of an overall website grading method that would quantify the quality and visits of a given site before and after reorganization, justifying thus the instantiation of certain redesign approaches.

Acknowledgments

The work of Yannis Panagis and Evangelos Sakkopoulos is partially supported by the Caratheodory research programme at the University of Patras. Yannis Panagis is also like to thank the European Social Fund (ESF), Operational Program for Educational and Vocational Training II (EPEAEK II) and particularly the Program PYTHAGORAS, for funding part of the above work.

References

1. D. Avramouli, J. Garofalakis, D.J. Kavvadias, C. Makris, Y. Panagis, E. Sakk-opoulos, "Popular Web Hot Spots Identification and Visualization", in the Four-teenth International World Wide Web Conference 2005 (WWW2005), Posters track, May 10–14, 2005, Chiba, Japan, pp. 912–913.
2. Boston Consulting Group, "Online Shopping Promises Consumers More than It Delivers", Boston Consulting Group Study, 2000.
3. M.-S. Chen, J.S. Park, and P.S. Yu. Efficient Data mining for path traversal patterns. *IEEE Trans. on Knowledge and Data Eng.*, 10(2), pp. 209–221, 1998.
4. Eleni Christopoulou, John Garofalakis, Christos Makris, Yannis Panagis, Evan-gelos Sakkopoulos, Athanasios Tsakalidis "Techniques and Metrics for Improv-ing Website Structure", Journal of Web Engineering, Rinton Press, 2,1–2 pp. 09–104, 2003.
5. Eleni Christopoulou, John Garofalakis, Christos Makris, Yannis Panagis, Evan-gelos Sakkopoulos, Athanasios Tsakalidis, "Automating Restructuring of Web Applications", ACM Hypertext 2002, June 11–15, 2002, College Park, Mary-land, USA., ACM 1-58113-477-0/02/0006.
6. R. Cooley. Web Usage Mining: Discovery and Application of Interesting Patterns from Web data. PhD thesis, University of Minnesota, 2000.
7. Drott M.C. Using web server logs to improve site design *Proceedings of ACM SIGDOC 98* pp. 43–50, 1998.
8. Garofalakis, J.D., Kappos, P. & Mourloukos, D.: Web Site Optimization Using Page Popularity. *IEEE Internet Computing* 3(4): 22–29 (1999)
9. John Garofalakis, Evangelos Sakkopoulos, Spiros Sirmakessis, Athanasios Tsakalidis "Integrating Adaptive Techniques into Virtual University Learning Environment", IEEE International Conference on Advanced Learning Technolo-gies, Full Paper, September 9–12, 2002, Kazan Tatarstan, Russia.
10. D.E. Knuth, Optimum Binary Search Trees. *Acta Informatica*, 1, 14–25, 1973.
11. K. Mehlhorn, Sorting and Searching. Data Structures and Algorithms, Vol. 1. EATCS Monographs in Theoretical Computer Science, Springer Verlag, 1984.
12. D.D. Sleator, and R.E. Tarjan. Self-Adjusting Binary Search Trees. Journal of the ACM, 32:3, 652–686, 1985.
13. R. Srikant, Y. Yang, Mining Web Logs to Improve Web Site Organization, in Proc. WWW01, pp. 430–437, 2001.

A Multi-Layered and Multi-Faceted Framework for Mining Evolving Web Clickstreams

Olfa Nasraoui

Department of Computer Science and Engineering, University of Louisville
olfa.nasraoui@louisville.edu

Abstract. Data on the Web is noisy, huge, and dynamic. This poses enormous challenges to most data mining techniques that try to extract patterns from this data. While scalable data mining methods are expected to cope with the size challenge, coping with evolving trends in noisy data in a continuous fashion, and without any unnecessary stoppages and reconfigurations is still an open challenge. This dynamic and single pass setting can be cast within the framework of mining evolving data streams. Furthermore, the heterogeneity of the Web has required Web-based applications to more effectively integrate a variety of types of data across multiple channels and from different sources such as content, structure, and more recently, semantics. Most existing Web mining and personalization methods are limited to working at the level described to be the lowest and most primitive level, namely discovering models of the user profiles from the input data stream. However, in order to improve *understanding* of the *real intention and dynamics of Web clickstreams*, we need to extend reasoning and discovery *beyond* the usual data stream level. We propose a new multi-level framework for Web usage mining and personalization, consisting of knowledge discovery at different granularities: *(i) session*/user *clicks, profiles, (ii) profile life events and profile communit*ies, and *(iii) sequential patterns* and *predicted shifts in the user profiles.* One of the most promising features of the proposed framework address the challenging dynamic scenarios, including *(i)* defining and detecting events in the life of a synopsis profile, such as *Birth, Death* and *Atavism,* and *(ii)* identifying Node *Communities* that can later be used to track the temporal evolution of Web profile activity events and dynamic trends within communities, such as *Expansion, Shrinking,* and *Drift.*

1 Introduction

(Everything flows, nothing is stationary) ***HERACLITUS, c.535–475 BC***
 The Web information age has brought a dramatic increase in the sheer amount of information (Web content), the access to this information (Web usage), as well as the intricate complexities governing the relationships within this information (Web structure). One of the side effects of this multi-faceted expansion is information overload, when searching and browsing websites. One

O. Nasraoui: *A Multi-Layered and Multi-Faceted Framework for Mining Evolving Web Click-streams,* Studies in Computational Intelligence (SCI) **14,** 11–35 (2006)
www.springerlink.com   © Springer-Verlag Berlin Heidelberg 2006

of the most promising and potent remedies against information overload comes in the form of personalization. *Personalization* aims to customize the interactions on a website depending on the users' explicit and/or implicit interests and desires. To date the most advanced and successful Web personalization approach has been based on *data mining* to automatically and efficiently uncover hidden patterns in Web data, such as is typically generated on busy Websites. Mining Web data is frequently referred to as *Web mining*, and it offers some of the most promising techniques to analyze data generated on a Website in order to help understand how users navigate through a given website, what information appeals to their variety of interests, and what peculiar information needs drive them in their browsing sessions. In addition to this understanding, Web mining can be used to improve a Website design and to provide automated and intelligent personalization that tailors a user's interaction with the website based on the user's interests. Understanding Web users' browsing patterns and personalizing their web navigation experience is beneficial to all users, but it is particularly more crucial on websites that are visited by a large variety of visitors with varying levels of expertise and background knowledge, and with distinct information needs and interests.

We propose a new framework for learning synopses in evolving data streams. The synopses are based on hybrid learning strategies that combine the power of distributed learning with the robustness and speed of statistical and mathematical analysis. While the theoretical foundation of the proposed methods are generic enough to be applied in a variety of dynamic learning contexts, we will focus on adaptive Web analytics and personalization systems that can handle massive amounts of data while being able to model and adapt to rapid changes, and while taking into account multi-faceted aspects of the data to enrich such applications. In particular, we propose a new multi-level framework for Web usage mining and personalization that is depicted in Fig. 1, consisting of knowledge discovery at different granularities: *session*/user *clicks, profiles, profile life events and profile community*, and *sequential patterns* and *predicted shifts in the user profiles*. The most important features of the proposed system are summarized below:

1. **Mining Evolving Data Streams:** Data is presented in a stream, and is processed sequentially as it arrives, in a single pass over the data stream. A *stream synopsis* is learned in a continuous fashion. The stream synopsis consists of a set of *dynamic synopsis nodes or clusters* that offer a summary of the data stream that is concise, yet semantically rich in information, as explained for Web data in item No. 2 below.

2. **Mining Evolving Web Clickstreams:** Applying the mechanics of mining evolving data streams to the web clickstream scenario to continuously discover an evolving profile synopsis, consisting of synopsis nodes. Each synopsis node is an entity summarizing a basic web usage trend that is characterized by the following descriptors: *typical representative user session summary, spatial scale, temporal scale, velocity vector, age*, and *mass*

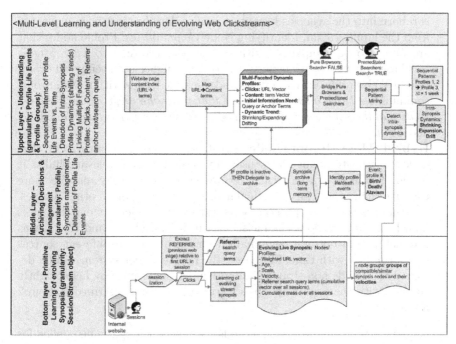

Fig. 1. Overall Learning and Understanding System for Mining Evolving Web Clickstreams

(number of objects from the input data stream falling within its influence zone based on a *Chebyshev Hypothesis test*).

3. **Integrating Multiple Information Channels:** In addition to web clicks, including information about web page content such as the "title", and also including information about the user's "Initial Information Need" from the "REFERRER" field in Web logs that indicates where the user came from before visiting the current website (Most users visit a website following a query in a search engine such as Google). Explicit queries tell us about the original *information need* that led the user to the current website. Even users with no such explicit information need could be linked to other users who happen to have the same browsing trend. Hence explicit information need in some users can elicit an implicit information need in other similar users. This in turn can serve as an additional information channel to support personalization and Web user understanding.

4. **Higher-level Web Analytics:** Defining and detecting events in the life of a synopsis profile, such as *(i)* *Birth*: a new profile is created when a never before seen object is presented from the input data stream, *(ii)* *Death*: No similar data is presented for a long time from the input stream that is similar to a previously discovered profile synopsis node. It is therefore placed in long term/secondary memory, and *(iii)* *Atavism*: a previously discovered profile synopsis node that went through a *Death* event,

is reborn into the synopsis because a similar data has just been presented from the input stream. Identifying Node *Communities*: Profile nodes that are found to be compatible with each other according to a hypothesis test based on *Chebyshev bounds*. Discovering higher level temporal evolution of Web profile activity events and dynamic trends within Web profile communities *based on intuitive rules that relate the angle between the velocities of the synopsis nodes in a node community: Expansion, Shrinking,* and *Drift*.

5. **Hybrid Meta-Recommender System:** We propose a system that makes a website adapt to its users by recommending relevant links. It is conceivable that a particular recommendation strategy may be more effective than others for a particular profile, but not for all other profiles. For these reasons, we propose a hybrid recommender system that uses knowledge from various channels listed in item 4 above, and that learns a dynamic and optimal combination strategy of these different sources of information with time. Moreover, the recommender system will be designed to provide non-stop, self-adaptive personalization by being tightly coupled with the stream synopsis learning procedure described in Item 2 above.

We emphasize that the proposed methods are generic enough to be applied for adaptive and continuous data summarization in a single pass (data stream) framework in a variety of applications. But we focus on the following application that arises when the data stream is a web clickstream, such as recorded in quickly expanding web user access logs on websites. An end tool of our research will be an intelligent Web usage analysis and Personalization system that can continuously adapt to changing content and user interests on a website.

2 Related Work

2.1 Mining Evolving Data Streams

Clustering [38] is an essential task of data mining that aims to discover the underlying structure of a set of data points, such as by partitioning the data into groups of similar objects. The accumulation of huge databases and repositories have made it impossible to understand the data in traditional interactive ways. This has resulted in an urgent need to sift through these huge data sets to discover knowledge quickly and automatically, thus the birth of data mining [17]. In the last decade, the continuous growth in the amounts of data stored in various repositories has placed even higher demands on clustering algorithms. They now must handle very large data sets, leading to some scalable clustering techniques [8, 48]. More recently, an explosion of applications generating and analyzing *data streams* has added new unprecedented challenges for clustering algorithms if they are to be able to track changing clusters in noisy data streams using only the new data points because storing past data is not even an option [3, 4, 6, 12, 13, 18, 28]. *Data streams* are massive data sets that arrive with a throughput that is so high that the data can only be analyzed sequentially and in a single pass. The discovery of useful patterns from these data streams is referred to as stream data mining. Because data streams unleash data points or measurements in a non-arbitrary order, they are inherently attached to a temporal aspect, meaning that the patterns that could be discovered from them follow dynamic trends, and hence they are different from traditional static data sets that are very large. Such data streams are referred to as *evolving data streams*. For these reasons, even techniques that are scalable for huge data sets may not be the answer for mining evolving data streams, because these techniques always strive to work on the entire data set without making any distinction between new data and old data, and hence cannot be expected to handle the notion of emerging and obsolete patterns. Like their non-stream counterparts, data streams are not immune from noise and outliers, which are data points that deviate from the trend set by the majority of the remaining data points. However, being able to handle outliers while tracking evolving patterns can be a tricky requirement that adds an additional burden to the stream mining task. This is because, at least the first time that an outlying data point is detected, it is not easy to distinguish it from the beginning of a new or emerging pattern.

Most clustering techniques such as K Means are based on the minimization of a Least Squares objective. The ordinary Least Squares (LS) method to estimate parameters is not robust because its objective function, $\sum_{j=1}^{N} d_j^2$, increases indefinitely with the residuals d_j between the jth data point and the estimated fit, with N being the total number of data points in a data set. Hence, extreme outliers with arbitrarily large residuals can have an infinitely large influence on the resulting estimate [21, 39]. All clustering techniques that are based on LS optimization, such as K Means, BIRCH [48], and Scalable K Means [8], inherit LS's sensitivity to noise.

There have been several clustering algorithms that were designed to achieve scalability [8, 48] by processing the data points in an incremental manner or in small batches. However these algorithms still treat all the objects of the data set the same way without making any distinction between old data and new data. Therefore, these approaches cannot possibly handle *evolving* data, where new clusters emerge, old clusters die out, and existing clusters change. More recently, several algorithms have been proposed and discussed more specifically within the context of stream mining [2, 6, 18]. STREAM [18] strives to find an approximate solution that is guaranteed to be no worse than a number of times the optimal. The optimal solution is based on minimizing the Sum of Squared Distances (SSQ), which is the same as the one used in K Means and LS estimates. However, LS estimates are not robust to outliers. Also, STREAMS finds a solution that approximates the entire data stream from beginning to end without any distinction between old data and newer data. Hence it has no provision for tracking *evolving* data streams. Fractal Clustering [6] defines clusters as sets of points with high self-similarity, and assigns new points in clusters in which they have minimal *fractal impact*. Fractal impact is measured in terms of the *fractal dimension*, and this can be shown to be related to the scale parameter. CluStream [2] is a recent stream clustering approach that performs *Micro-Clustering* with the new concept of *pyramidal timeframes*. This framework allows the exploration of a stream over different time windows, hence providing a better understanding of the evolution of a stream. the main idea in CluStream is to divide the clustering process into an online process that periodically stores summary statistics, and an offline process that uses only these summary statistics. Micro-clusters are an extension of BIRCH's Cluster Feature (CF) [48] with temporal statistics, and the incremental updating of the CF is similar to BIRCH. BIRCH, in turn, solves a LS criterion because the first order statistics are nothing more than the mean centroid values. For this reason, CluStream was not designed to handle outliers. In [28], we proposed a new immune system inspired approach for clustering noisy multi-dimensional stream data, called TECNO-STREAMS (Tracking Evolving Clusters in NOisy Streams), a scalable clustering methodology that gleams inspiration from the natural immune system to be able to continuously learn and adapt to new incoming patterns by detecting an unknown number of clusters in evolving noisy data in a single pass. TECNO-STREAMS relies on an evolutionary process inspired by by the natural immune system. Unfortunately, TECNO-STREAMS relied on many parameters in order to yield good results. TECNO-STREAMS was also *limited to work only at the data to stream synopsis level*, never going beyond that to infer any kinds of dynamic properties or rich temporal trend information that may be occuring within the synopsis.

The stream data mining approach proposed in this chapter differs from existing techniques in more than one way: *(i)* a robust clustering process is used to resist outliers, *(ii)* the number of clusters is not required to be known in advance, *(iii)* the goal of stream clustering is to form a continuously (non-stoppable) evolving synopsis or summary of the data stream, while focusing more on the *more recent data, (iv)* higher level semantics are discovered to capture the dynamics of communities within the stream synopsis such as expansion and drift, and temporal evolution of the synopsis trends.

2.2 Mining Evolving Web Clickstreams

(We are what we repeatedly do) **Aristotle, c.384 BC–322 BC**

In addition to its ever-expanding size and dynamic nature, the World Wide Web has not been responsive to user preferences and interests. *Personalization* deals with tailoring a user's interaction with the Web information space based on information about him/her, in the same way that a reference librarian uses background knowledge about a person or *context* in order to help them better. The concept of *contexts* can be mapped to distinct user *profiles*.

Manually entered profiles have raised serious *privacy* concerns, are *subjective*, and *do not adapt* to the users' changing interests. *Mass profiling*, on the other hand, is based on general trends of usage patterns (thus protecting privacy) compiled from all users on a site, and can be achieved by mining or discovering user profiles (i.e., clusters of similar user access patterns) from the historical *web clickstream* data stored in server access logs. The simplest type of personalization system can suggest relevent URLs or links to a user based on the user's interest as inferred from their recent URL requests. A *web clickstream* is a virtual trail that a user leaves behind while surfing the Internet, such as a record of every page of a Web site that the user visits. Recently, data mining techniques have been applied to discover mass usage patterns or profiles from Web log data [1, 7, 9, 14, 25, 23, 29, 32, 36, 37, 41, 42, 43, 44, 45, 46, 47]. Most of these efforts have proposed using various data mining or machine learning techniques to model and understand Web user activity. In [45], clustering was used to segment user sessions into clusters or profiles that can later form the basis for personalization. In [37], the notion of an adaptive website was proposed, where the user's access pattern can be used to automatically synthesize index pages. the work in [15] is based on using association rule discovery as the basis for modeling web user activity, while the approach proposed in [7] used Markov Random Fields to model Web navigation patterns for the purpose of prediction. The work in [47] proposed building data cubes from Web log data, and later applying Online Analytical Processing (OLAP) and data mining on the cube model. Spiliopoulou et al. [42] presents a complete Web Usage Mining (WUM) system that extracts patterns from Web log data with a variety of data mining techniques. In [31, 32], we have proposed new robust and fuzzy relational clustering techniques that allow Web usage clusters to overlap, and that can detect and handle outliers in the data set. A new subjective similarity measure between two Web sessions, that captures the organization of a Web site, was also presented as well as a new mathematical model for "robust" Web user profiles [32] and quantitative evaluation means for their validation. In [29], a *linear* complexity Evolutionary Computation technique, called Hierarchical Unsupervised Niche Clustering (H-UNC), was presented for mining both user profile clusters and URL associations. A density based evolutionary clustering technique is proposed to discover multi-resolution and robust user profiles in [30]. The K Means algorithm was used in [41] to segment user sequences into different clusters. An extensive survey of different approaches to Web usage mining can be found in [43].

Unfortunately, all the above methods assume that the entire preprocessed Web session data is static and could reside in main memory. Automatic Web personalization can analyze the data to compute recommendations in different ways, including: **(1) Content-based or Item-based filtering** This system recommends items deemed to be similar to the items that the user liked in the past. Item similarity is typically based on domain specific item attributes [5] (such as author and subject for book items, artist and genre for music items, or terms for web pages), **(2) Collaborative filtering** Based on the assumption that users with similar past behaviors (rating, browsing, or purchase history) have similar interests, this system recommends items that are liked by other users with similar interests [22, 26, 33, 34]. This approach relies on a historic record of all user interests such as can be inferred from their browsing activity on a website. **(3) Hybrid systems** tend to combine the strength of several familie of recommender systems, such as collaboartive and content-based filtering [11, 35].

3 Proposed Framework

3.1 Mining Evolving Data Streams

(You cannot step twice into the same stream. For as you are stepping in, other waters are ever flowing on to you) **HERACLITUS, c.535–475 BC, quoted by Plato**

In this section, we present a new paradigm for mining evolving data streams, that is based purely on mathematical and statistical concepts that are simple to understand, analyze and control; and also that goes far beyond the stream description level to capture higher level *community* formations wihin the stream synopsis, as well as *life cycle events* such as birth, death, and atavism of certain stream synopsis nodes, and *dynamic trends* within the synopsis such as phenomena related to *expansion, focusing,* and *drifting* within stream synopsis communities.

In a dynamic environment, the data from a data stream $\mathbf{X}_a$ are presented to the cluster model one at a time, with the cluster centroid and scale measures re-updated with each presentation. It is more convenient to think of the data index, j, as monotonically increasing with time. That is, the data points are presented in the following chronological order: $\mathbf{x}_1, \cdots, \mathbf{x}_N$. Hence after encountering J points from the data stream, a cluster is characterized by its location or center $\mathbf{c}_{i,J}$, its scale $\sigma_{i,J}^2$, and its age t_i which, to make independent of any time units, can be set to the number of points that have been streaming since the cluster's conception at $t_i = 0$. The set of clusters and their characteristic parameters define a *synopsis* [20] or good summary representation of the data stream, denoted as $\mathbf{B}$. As we will see below, because the currency of the stream is taken into account to define the influence zone around each cluster, the synopsis will also reflect a more current summary of the data stream, that will evolve with the changes in the data stream. In other words, there is never a final commitment to any cluster: Clusters are conceived when the possibility of a new dense area emerges from the stream (note that this includes possible outliers, because they cannot be discerned at the beginning) and clusters die out when they become so old and outdated that they no longer represent the current stream. In order to adhere to the memory requirements of a stream scenario, the maximal number of clusters is fixed to an upper bound that depends on the maximum size alloted for the stream synopsis. Each candidate cluster defines an influence zone over the data space. However, since data is dynamic in nature, and has a temporal aspect, data that is more current will have higher influence compared to data that is less current. The influence zone is defined in terms of a weight function that decreases not only with distance from the data to the cluster prototype, but also with the time since the data has been presented to the cluster model. It is convenient to think of time as an additional dimension to allow the presence of evolving clusters.

Definition 1 (Adaptive Robust Weight)

For the ith cluster, $\mathbf{C}_i$, $i = 1, \cdots, C$, we define the robust weight of the jth data point, at the moment when the total size of the data stream accumulated to J inputs: $\mathbf{x}_1, \mathbf{x}_2, \cdots, \mathbf{x}_j, \cdots, \mathbf{x}_J$, as

$$w_{ij,J} = w_{i,J}\left(d_{ij}^2\right) = e^{-\left(\frac{d_{ij}^2}{2\sigma_{i,J}^2} + \frac{(J-j)}{\tau}\right)} \tag{1}$$

where τ is an application-dependent parameter that controls the time decay rate of the contribution from old data points, and hence how much emphasis is placed on the currency of the cluster model compared to the sequence of data points encountered so far. d_{ij}^2 is the distance from data point $\mathbf{x}_j$ to cluster location $\mathbf{c}_{i,J}$. $\sigma_{i,J}^2$ is a *scale or dispersion* parameter that controls the decay rate of the weights along the spatial dimensions, and hence defines the size of an influence zone around a cluster prototype. Data samples falling far from this zone are considered outliers. At any point J in the stream sequence, the weight function of data point $\mathbf{x}_j$ in cluster $\mathbf{C}_i$ decreases geometrically with the age $(t_j = J - j)$ or number of samples encountered since $\mathbf{x}_j$ was introduced. Therefore the weights will favor more current data in the learning process.

At any point J in the stream (after encountering J data points), we search for the optimal *dense* cluster locations $\mathbf{c}_{i,J}$ and scale values $\sigma_{i,J}^2$, by optimizing the following criterion

$$\max_{\mathbf{c}_{i,J}, \sigma_{i,J}} \left\{ \mathcal{J}_{i,J} = \sum_{j=1}^{J} \frac{w_{ij,J}}{\sigma_{i,J}^2} \right\}, \quad i = 1, \cdots, C, \tag{2}$$

Because the weight $w_{ij,J}$ can also be considered as the degree of membership of data point $\mathbf{x}_j$ in the *inlier* set or the set of good points, the numerator of the above criterion consists of a soft estimate of the *cardinality* (sum of weights) of the inlier (non-outliers) set. Hence, by maximizing this cardinality, the objective function tries to use as many good points (inliers) as possible in the estimation process, via their high weights, so that efficiency is least compromised. At the same time, trivial solutions corresponding to infinite scale or inlier boundaries (and thus would include all the data) are penalized by the denominator. Thus the combined effect is to optimize the *density*, i. e., the ratio of the total number of good points to the scale. Finally, we should note that d_{ij}^2 should be a suitable distance measure, tailored to detect desired shapes, such as the Euclidean distance for spherical clusters. For the case, where the distance measure is differentiable with respect to the cluster representatives, analytical optimization is possible based on iterative alternating optimization, where a set of parameters are fixed while the remaining parameter is updated, such as used in the Expectation Maximization (EM) approach [16]. For other arbitrary and non-differentiable dissimilarity measures, a different optimization method may be adopted (*bootstrapping*). We begin with the case of differentiable dissimilarity measures, typical with applications dealing with interval scaled data attributes. Since the objective function depends on several variables, we can use an alternating optimization technique, where in each iteration a set of variables is optimized while fixing all others.

Lemma 1. *Optimal Incremental Center Update*

Given the previous centers resulting from the past $(J-1)$ data points, $\mathbf{c}_{i,J-1}$, the new centroids that maximize (2) after the Jth *non-outlier (following a Chebyshev test)* data point is given by

$$\mathbf{c}_{i,J} = \frac{e^{-\frac{1}{\tau}}\mathbf{c}_{i,J-1}W_{i,J-1} + w_{iJ,J}\mathbf{x}_J}{\left(e^{-\frac{1}{\tau}}W_{i,J-1} + w_{iJ,J}\right)} . \tag{3}$$

where $W_{i,J-1} = \sum_{j=1}^{J-1} w_{ij,(J-1)} = W_{i,J-2} + w_{i(J-1),(J-1)}$ is the *mass* or sum of the contributions from previous data points in this cluster/node, $\mathbf{x}_1, \mathbf{x}_2, \cdots, \mathbf{x}_{J-1}$.

Idea of the Proof: Since the time dependency has been absorbed into the weight function, and by fixing the previous centroids, $\mathbf{c}_{i,J-1}$, scale $\sigma_{i,J-1}^2$ and weight sums, $W_{i,J-1}$, the equations for center updates are found by solving

$$\frac{\partial \mathcal{J}_{i,J}}{\partial \mathbf{c}_{i,J}} = \frac{1}{\sigma_{i,J}^2} \sum_{j=1}^{J} w_{ij,J} \frac{\partial d_{ij}^2}{\partial \mathbf{c}_{i,J}} = 0 .$$

Each term that takes part in the computation of $\mathbf{c}_{i,J}$ is updated individually with the arrival of each new data point using the old values, and adding the contribution of the new data sample. Note also that subsequently to each new point that is read from the stream, and assuming that the scale parameter does not change too much as a result of a single input (i.e. $\sigma_{i,J}^2 = \sigma_{i,J-1}^2$), each weight decays as follows to enable the forgetting process, and also allows us to extract the previous centers $\mathbf{c}_{i,(J-1)}$:

$$w_{ij,J} = e^{\frac{-1}{\tau}} w_{ij,(J-1)} . \tag{4}$$

Lemma 2. *Optimal Incremental Scale Update to maximize (2)*

$$\sigma_{i,J}^2 = \frac{2e^{-\frac{1}{\tau}}\sigma_{i,J-1}^2 W_{i,J-1} + w_{ij,J}d_{iJ}^2}{2\left(e^{-\frac{1}{\tau}}W_{i,J-1} + w_{iJ,J}\right)}. \tag{5}$$

Idea of the Proof: Since the time dependency has been absorbed into the weight function, and by fixing the previous centroids, $c_{i,J-1}$, scale $\sigma_{i,J-1}^2$, and weight sums, $W_{i,J-1}$, the equations for scale updates are found by solving $\frac{\partial \mathcal{J}_{i,J}}{\partial \sigma_{i,J}^2} = \mathbf{0}.$

Therefore, the algorithm for incrementally adjusting the optimal cluster locations and scales will consist of updates of the prototype parameters, followed by updates of the scale parameter and the weights in an iterative fashion, with the arrival of each new data point in the stream. Note that only the cumulative statistics need to be stored for each cluster at any given instant. Hence the synopsis consists of a summary of all the clusters, where each cluster is represented by the tuple consisting of cluster representative $c_{i,J}$, scale or inlier boundary $\sigma_{i,J}^2$, and the total mass (sum of individual weight contributions from the previous data stream) $W_{i,J}$. These are all that is needed to perform each incremental updating.

Learning New Data Points and Relation to Outlier Detection

Definition 2. *Potential Outlier*

A *potential outlier* is a data point that fails the outlyingness test for the entire cluster model. The outlier is termed *potential* because, initially, it may either be an outlier or a new emerging pattern. It is only through the continuous learning process that lies ahead, that the fate of this outlier will be decided. If it is indeed a true outlier, then it will form no mature clusters in the cluster model.

Chebyshev Bounds

Several upper tail bounds exist in statistics that bound the total probability that some random variable is in the tail of the distribution, i.e., far from the mean. Markov bounds apply to any non-negative random variable, and hence do not depend on any knowledge of the distribution. However, a *tighter bound can be obtained using Chebyshev bounds* if a reliable estimate of scale is available [19]. Again, *no assumptions are made about the distribution* of the data, other than scale, of which we have robust estimates. The Chebyshev bound for a random variable X with standard deviation σ and mean value μ is:

$$Pr\left\{|X - \mu| \geq t\sigma\right\} \leq \frac{1}{t^2} \tag{6}$$

Testing a Data Point or a New Cluster for Outlyingness with Respect to Cluster C_i Using Chebyshev Bound with Significance Probability $1/t^2$

(6) can be rearranged in the form

$$Pr\left\{|X-\mu|^2 \geq t^2\sigma^2\right\} \leq \frac{1}{t^2}, \text{ or equivalently } Pr\left\{e^{\frac{-|X-\mu|^2}{2\sigma^2}} \leq e^{-t^2/2}\right\} \leq \frac{1}{t^2}$$

This results in the following test for outlier detection:

Chebyshev Test

IF $w_{ij,J} < e^{(-t^2/2)}$ THEN x_j is an outlier with respect to cluster C_i

The same test is used in the algorithm to decide when a new cluster is created, since points from new clusters can be considered as outliers with respect to old clusters. Note that the quantity $t^2\sigma^2$ is referred as the *Chebyshev Bound* from now on, and it is considered as the distance from the centers that includes all the inliers with probability $1 - \frac{1}{t^2}$.

Extending the Density Criterion into the Time Domain

In the *synopsis node* model, the weights w_{ij} decay in both spatial domain and time domain. While the decay in the spatial domain is governed by an optimal spatial scale parameter σ_i^2, the temporal decay is governed by a time constant τ. While the optimal spatial scale parameter σ_i^2 corresponds to the boundary that ensures an optimal soft density within its confines, the time constant τ is expected to depend on the pace or time characteristics of the dynamic data. This leads us to consider whether the idea of density based scale value in the spatial domain could be extended to a density based scale value in the temporal domain. Once we exploit this analogy, it is natural to think of specializing the temporal scale to the different synopsis nodes, thereby, leading to a more refined synopsis node model that would be able to continuously capture both the spatial and temporal charateristics of each pattern. Hence, for the ith synopsis node, we redefine the weight/membership function as follows. $w_{ij} = w_i\left(d_{ij}^2\right) = e^{-\left(\frac{d_{ij}^2}{2\sigma_i^2} + \frac{j}{\tau_i}\right)}$ The density criterion, after J input data have been presented, is extended to a spatio-temporal density: $\mathcal{J}_{i,j} = \frac{\sum_{j=1}^{J} w_{ij}}{\sigma_i^2 \tau_i}$. By setting $\frac{\partial \mathcal{J}_{i,j}}{\partial \sigma_i^2} = 0$, the iterative Picard update equations for spatial scale do not change. Similarly, setting $\frac{\partial \mathcal{J}_{i,j}}{\partial \tau_i} = 0$, yields iterative Piccard update equations for the temporal scale:

$$\tau_i = \frac{\sum_{j=1}^{J} w_{ij} j}{\sum_{j=1}^{J} w_{ij}} \tag{7}$$

Continuous Learning by Bootstrapping the Synopsis

When the distance measure is non-differentiable, we need an alternative way to estimate the optimal centers. One way to do this is through bootstrapping the stream synopsis, i.e., resampling from the synopsis nodes, and hence creating duplicates of the good synopsis nodes. More specifically, we will use a more competitive form of bootstrapping to provide an incentive for better synopsis nodes to dominate over bad ones in the long run, hence enabling a continuous and incremental latent optimization process that is particularly useful for applications with non-differentiable criteria. This *bootstrapping* variant is based on *resampling based on a boosting or arcing strategy* [40] that promotes better solutions in sampling to improve the current synopsis. To further understand the latter analogy with arcing, consider the way arcing is applied to improve a sequence of R classifiers for m samples. The probability of inclusion of the ith data point at the $r + 1$ step for the $(r + 1)$th classifier can for example be given by $p^{r+1}(i) = \frac{1+k(i)^4}{\sum_{j=1}^{m} 1+k(j)^4}$, where $k(i)$ is the number of misclassifications of the ith point. Instead of classification, our goal here is to form a reasonable *summarization model* or synopsis of the data stream as represented by the set of nodes in the current network. Note that this synopsis is the only view/summary that we have of the data stream. Alternately, classification can be replaced by anomaly detection, a dual of the summarization goal, since anomalous data is an instance of data that should *not be* summarized. Hence, in each step (after encountering each datum), we attempt to improve the current synopsis using the only possible arcing strategy available, which is by resampling from the current set of synopsis nodes that model the data stream. In our context, instead of a number of misclassifications $k(i)$, we use the density criterion function to compute the inclusion probabilities that will define the synopsis node summarization model of the current data stream for the next step. This results in an arcing/bootstrapping probablity that is a function of the density/goodness of the $i$thsynopsis node, such as given by $P_{arced_i} = \frac{\mathcal{J}_i}{\sum_{k=1}^{C} \mathcal{J}_k}$.

Modeling Temporal Change via Velocities

When a new datum $\mathbf{x}_J$ from the data stream does not fail the Chebychev test, it is considered as a non-outlier or a data point that is already represented in the stream synopsis by a node (this is the closest node to the new data point in terms of the dissimilarity measure). Currently, the effect of this new data point is to possibly affect this closest representative such as by incremental update of the centroid location and scale, as well as the age. In order to model the dynamics of the change in location, we further compute the velocity vector of the current node, and add it to the information stored in this node, as follows

$$\mathbf{v}_{i,J} = \frac{\mathbf{c}_{i,J} - \mathbf{c}_{i,J-1}}{t_i}. \tag{8}$$

Note that the velocity of a node measures the amount of change in the node location per unit time, where the denominator measures the time since the last update of this node, which corresponds to the effective age of the node (the time since it was last activated by a data point).

Complete Algorithm for Bootstrapping Synopsis to Mine Evolving Data Streams

Finally we give the steps of an algorithm to implement all the steps explained in this section below:

Fix the maximal number of cluster prototypes, C_{max};
Initialize centroids $\mathbf{c}_{i,J} = \mathbf{x}_i$, $i = 1, \cdots, C_{max}$;
Initialize scale $\sigma_{i,J} = \sigma_0$, $i = 1, \cdots, C_{max}$;
Initialize age $t_i = 0$, $i = 1, \cdots, C_{max}$;
Initialize the sums $W_{i,J} = 0$, $i = 1, \cdots, C_{max}$;
Initialize $C = C_{max}$;
Let stream synopsis model $B = C_1 \cup C_2 \cup \cdots \cup C_C$, and synopsis node representatives after J data points, be $\mathbf{C}_i = (\mathbf{c}_{i,J}, \sigma_{i,J}, t_i, W_{i,J}, \tau_i)$;
    FOR $J = 1$ **TO** N **DO** // single pass over the data stream $\mathbf{x}_J$
        FOR $i = 1$ **TO** C **DO**
        Compute distance, $d_{i,J}^2$, and robust weight, $w_{ij,J}$;
        IF $C < C_{max}$ AND $\mathbf{x}_J$ is an outlier (based on Chebyshev test) relative to all synopsis nodes C_i THEN {
           Create new synopsis node C_k: $B \leftarrow B \cup C_k$ with $k = C + 1$;
           $\mathbf{c}_k = \mathbf{x}_j$;
           $\sigma_k = \sigma_0$;
           Initialize $t_k = W_{k,J} = 0$;
           $C = C + 1$
        }
        Perform Bootstrapping on the set of synopsis nodes using arcing strategy;
        FOR $i = 1$ **TO** C **DO**
        IF $\mathbf{x}_J$ is not an outlier in $\mathbf{C}_i$ (based on Chebyshev test) **THEN**{
           Update $\sigma_{i,J}^2$ using (5);
           Update $\tau_{i,J}$ using (7);
           Update $\mathbf{c}_{i,J}$ using (3)
           Update $W_{i,J} = W_{i,J-1} + w_{ij,J}$;
           Update age $t_i = t_i + 1$; // age since inception
           Update velocity v_i using (8);
         }
        FOR $i = 1$ **TO** C **DO**
        IF $\left(\mathcal{J}_{(i)} < \mathcal{J}_{min} \right)$ AND $t_i > t_{min}$ THEN {
           $B \leftarrow B - C_i$; // Eliminate nodes that are mature and weak, and move them to secondary memory
           $C = C - 1$;
        }
        Test nodes for compatibility to form communities of compatible synopsis nodes;
    }

Computational Complexity

Our proposed algorithm to mine evolving data streams necessitates the iterative computations of distances and weights for each data vector, followed by the center and scale parameter, but these computations are computed only once for each input from the data stream. These are all linear in the number of data vectors. Hence the computational complexity is $\mathcal{O}(N)$. At any point in time throughout the stream sequence, only the most recent data point $\mathbf{x}_J$

is needed to incrementally update all the cluster components. Therefore, the memory requirements are obviously linear with respect to the maximal number of synopsis nodes/clusters, C_{max}, which is only a negligible fraction of the size of the data set.

3.2 Mining Evolving Web Clickstreams and Non-Stop, Self-Adaptive Personalization

The Web personalization process can be divided into two components:

(1) Mining Evolving Web Clickstreams: As an alternative to locking the state of the Web access patterns in a frozen state depending on when the Web log data was collected and preprocessed, we propose an approach that considers the Web usage data as a reflection of a dynamic environment. We plan to use our proposed continuous synopsis learning approach to continuously and dynamically learn evolving Web access patterns from non-stationary Web usage environments. To summarize our approach: **(1)** The input data can be extracted from web log data (a record of all files/URLs accessed by users on a Web site), **(2)** The data is pre-processed (e.g. via cookies) to produce session lists: A session list $\mathbf{s}_i$ for user i is a list of URLs visited by the same user, represented as a binary vector (1: URL was visited, 0: URL was not visited), **(3)** The ith synopsis node represents the ith candidate profile $\mathbf{P}_i$ and encodes relevant URLs, which are the attributes in this case, as well as *scale* (spatial and temporal), *age, mass,* and *velocity.* The proposed approach to mining evolving data streams can learn a synopsis consisting of an unknown number of evolving profiles *in real time.* Hence, it is suitable for use in a real time personalization system to constantly and continuously provide the recommendation engine with a fresh and current list of user profiles. Old profiles can be handled in a variety of ways. They may either be discarded, moved to secondary storage, or cached for possible re-emergence. Even if discarded, older profiles that re-emerge later, would be re-learned like new profiles. Hence the logistics of maintaining old profiles are less crucial than with most existing techniques. Personalization can be achieved in the first stage by continuously discovering a proper set of possibly evolving usage profiles or contexts in real time (single pass). The discovered profiles will thus form an evolving knowledge base of usage contexts.

(2) Real Time Recommendation Engine: The last phase in personalization (see Fig. 2) makes use of the results of mining evolving web clickstreams (i.e. the continuously evolving profiles contained in the stream synopsis) to deliver recommendations to the user. The recommendation process typically involves generating dynamic Web content on the fly, such as adding hyperlinks to the last web page requested by the user. This can be accomplished using a variety of Web technology options such as CGI programming. Given a current unfinished user session, it is mapped to one or several pre-discovered usage profiles to form the recommendations, in a very similar manner to classification. The classes in this case will be based on the dynamically discovered

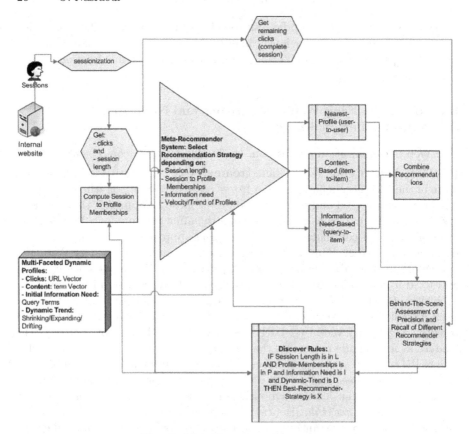

Fig. 2. Meta-Recommender system

profiles in the stream synopsis. Prediction will involve computing a set of URLs to be recommended. The simplest recommendation engine would first match a current user session to one of the profiles, and then recommend the URLs that make up the profile and that are not part of the current session. However, we propose more elaborate methods to

1. compute recommendations that can tap on different channels of data in addition to the usage (i.e. clickstream) data contained in web clickstreams. These include content, semantics, and REFERRER information, and are discussed in the next sections.
2. perform a hybrid recommendation strategy that combines the strengths of several different recommendation strategies, such as *(i)* content based filtering that suggests items similar to items in the current user session (where content is inferred from any or multiple information channels as explained in the previous item), *(ii)* collaborative filtering that recommends items in user profiles that are similar to the current user (here again we can use multi-faceted user profiles that combine different information channels:

clicks, title terms, REFERRER search query terms and inferred Information Need), and *(iii)* semantic recommendations that suggest items that are compatible with the ones in the current session according to available website ontologies.

3.3 Integrating Multiple Sources for Information Need Assessment

(From out of all the many particulars comes oneness, and out of oneness come all the many particulars) **HERACLITUS, c.535–475 BC**

Users generally access the pages of a website trying to satisfy a certain *information need.* The holy grail of Web personalization (and of search engines as well) is guessing the information need of the user. Based only on their clicks, it is hard to accurately estimate this information need. Therefore, we propose to gather the forces of other information channels to accomplish this task.

Step 1: Title and REFERRER Term Collection Starting with the Web logs that contain all URLs visited by users on a web site, suppose that a user's clicks are: *a, b, c.* Further suppose that we further look at the *REFERRER* field that includes the URL of the web page that the user visited just prior to visiting *a*, then we can reconstruct from the log file the session: *(REFERRER, a, b, c).* Most of the visits to Information portals tend to come from the results of a search query on a search engine such as Google. In this case, we extract from the *REFERRER* field the *search query terms* that the user typed when searching for information that led them to this website, this gives us some terms that describe the user's Information Need that motivated him/her to initiate the current session. In addition to knowledge about the Information Need of the user, we can also infer the terms that describe the topic of their interest during this session by extracting the *title terms* that are within the <TITLE> HTML tags in each clicked page. At the end of collecting all title terms, we end up with terms describing each URL in session (a,b,c). It is possible that the title terms cannot be extracted for special file types such as images and audio files. In this case we can resort to the <ALT> tags that give an annotation of the special file, and even if the <ALT> tag is not available, we can simply ignore the content of this file, because we plan to also use collaborative filtering (user-to-user recommendations) known to capture the meaning of special content files through their association with other regular files.

Step 2: Integrating Content and REFERRER Information into Web Usage Mining After transforming each user sequence of clicked URLs into a sequence of title terms, we can proceed to perform web usage mining in the *URL (click)* domain or *title term* domain, and extract the profiles as a set of URLs/title terms relevant to user groups. Hence we have *two different facets of the user profiles: clicks/URLs* and *title terms,* and *hence two different recommender strategies* to complement one another.

Step 3: Bridging Search Query Terms and User Profiles for Information Need Assessment To estimate the user's intitial information need or purpose of visiting the website, we need to consider the *REFERRER* field for the very first URL clicked in each session where the *REFERRER* includes a search query (such as from Google). Later on after the user profiles are discovered, the frequency of the search query terms are tallied incrementally to characterize each profile. This can be used to label even the sessions in the profile that do not come from a search engine (pure browsing based sessions) with search terms of similar sessions (i.e. in the same profile) that did get initialted from an explicit *search*, and hence help bridge the gap between

1. *Pure Browsers*: users who just visit this website *without* coming from a search engine, and
2. *Premeditated Searchers*: users who visit as a result of clicking on one of the results of a search query on a search engine.

This is one way to enrich our knowledge of the user's *Information Need*. In fact, the Information Need assessment goes both ways: knowing one, we can predict the other, i.e.

1. from *Pure Browsers' clicks or URLs*, we can predict the most likely *search terms* that would describe the *Information Need* for this user.
2. from the *search query terms* that would describe the *Information Need* of a *Premeditated Searcher* (i.e. from the first URL clicked into our website), predict the most likely pure browsing *clicks* that would be initiated by this user.

Later on, one possible application to recommendations is to suggest the URLs in profiles of previous users with similar search query terms. We call this *On-First-Click-Search-Mediated-Recommendations*: they directly bypass a lenghty navigation session by bridging the gap between pure browsers and premeditated searchers. Another possibility is to map search terms to the documents/content on website. This will transcend clicks/usage access data and map URLs to popular search terms.

3.4 Putting the Icing on the Cake: Higher-Level Web Analytics

(Much learning does not teach understanding) **HERACLITUS, c.535–475 BC**

Most existing Web mining and personalization methods are limited to working at the level described to be the lowest and most primitive level in our chapter, namely discovering models of the user profiles from the input data stream. However, as discussed below, in order to improve *understanding* of the *real intention behind clickstreams*, we need to extend reasoning and discovery *beyond* the usual data stream level.

We propose a new multi-level framework for Web usage mining and personalization that is depicted in Fig. 1.

1. The *lowest level* works on the session clickstream level to automatically form *a summary synopsis of the input clickstream* in the form of user profiles.
2. The *intermediate* level works *at the level of the discovered profiles* from the lowest level to define and detect life events such as *birth, death, and atavism* and also detects dynamic trends within synopsis profile communities such as *expansion, shrinking/focusing, and drifting*.
3. The *highest level* works at the level of *profile life events and profile community dynamic trends* to infer higher level knowledge such as *sequential patterns* and *predicted shifts in the user profiles*.

Extracting Dynamic Trends within Stream Synopsis Communities from Velocities

Forming Synopsis Communities by testing the compatibility of synopsis nodes $\mathbf{C}_i$ *and* $\mathbf{C}_k$ *with scales* $\sigma^2_{i,J}$ *and* $\sigma^2_{k,J}$ *using Mutual Chebyshev Bounds with Significance Probability* $1/t^2$:

Given the distance between these two clusters, d^2_{ik}, if the clusters are compatible based on mutual Chebyshev tests, then $\mathbf{C}_i$ and $\mathbf{C}_k$ are in the same community. The centroid for each community is computed by a continuous/incremental procedure as follows. Starting with being initialized to the first node that is part of the community, the centroid of the community of synopsis nodes is updated as follows each time that a new node $\mathbf{C}_k$ from the synopsis joins the community: $\mathbf{c}_{new,J} = \frac{c_{old}W_{old}+c_{k,J}W_{k,J}}{W_{old}+W_{k,J}}$. Forming communities within the stream synopsis allows us to extract rich information about the dynamics of the evolving data stream. For example, it is possible to discover dynamic trends within Web profile communities *based on intuitive rules that relate the angle between the velocities of the synopsis nodes in a node community: Expansion, Focusing, and Drift*. Each dynamic behavior can be detected based on the following simple rules, which are intuitive as can be verified in Fig. 3:

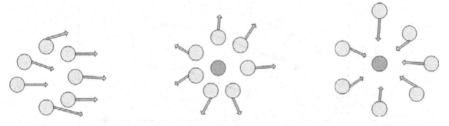

Fig. 3. Dynamic Trends within Synopsis Communities: Drifting (*left picture*), Expansion (*middle*), and Shrinking (*right*) clear circles = synopsis nodes, darker circle = centroid of community, arrows = velocities

- *Drifting*: The velocity angles within the commmunity are similar: their variance from the mean angle value is close to 0.
- *Expansion*: On average, the velocity angle for each node is at a difference of 180 degrees relative to the angle of the vector joining the node and the centroid of the community.
- *Focusing*: On average, the velocity angle for each node is at a difference of 0 degrees relative to the angle of the vector joining the node and the centroid of the community.

Learning the Temporal Evolution of User Profiles

(Immortal mortals, mortal immortals, one living the others death and dying the others life) **HERACLITUS, c.535–475 BC**

We define several event types in the life of a synopsis profile, such was *(i) Birth*: a new profile is created when a never before seen object is presented from the input data stream, *(ii) Death*: No similar data is presented for a long time from the input stream that is similar to a previously discovered profile synopsis node. It is therefore placed in long term/secondary memory, and *(iii) Atavism*: a previously discovered profile synopsis node that went through a *Death* event, is reborn into the synopsis because a similar data has just been presented from the input stream. These events are illustrated in Fig. 4. Later, we will detect and track the temporal dynamics of profiles through these pre-defined events: birth, death, and atavism (rebirth) by symbolic time series analysis or mining sequential episodes. This approach can be used to visualize the evolution of user interests, discover seasonality and other higher level causality or cross-temporal correlations. For example, *Sequential Association Rules* of the form: {Profile1 → Profile2: support= 5%, confidence = 50%, time difference = n days}.

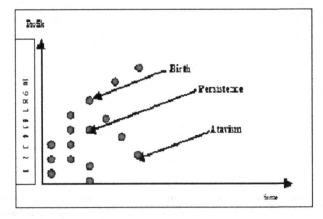

Fig. 4. *Profile Life Events* defined to track profile evolution

3.5 Hybrid Recommender System

An illustration of the proposed system is shown in Fig. 2. *Collaborative filtering* based methods recommend items based on collective *"usage"* patterns, by recommening items of interest to the users in the same neighborhood or similar to the current user. Hence, no item content information is used. This can cause a problem when new items/web pages are added for which there is no user information. This is called the *new item problem*. *Content-based filtering* methods can avoid the new item problem by recommending items with similar content to the items that the user showed interest in. However, content is very elusive for special types of items on the web such as most graphic items and audio files, and even for text because it is sometimes difficult to extract accurate semantics from free text. For these special file types, collaborative filtering is the best strategy because it only relies on past *usage* data not content, and hence is able to capture even latent, non obvious similarities between item content. Furthermore, it is conceivable that a particular recommendation strategy may be more effective than others for a particular profile, but not for all other profiles. For these reasons, we propose a hybrid recommender system [10] that uses knowledge from various channels described above, and that learns a dynamic and optimal combination strategy of these different sources of information with time. Moreover, the recommender system will be designed to provide non-stop, self-adaptive personalization by being tightly coupled with the evolving stream synopsis mining procedure. Knowledge about the optimal recommender system will be discovered in the form of rules such as (*IF profile is X and Information Need is Y THEN Optimal recommender = Z*) based on rules that are mined from continuous behind-the-scene recommender system validation (explained later in the validation section). Various temporal sequence mining methods can be used to this end [24].

4 Conclusion

Recently, an explosion of applications generating and analyzing *data streams* has added new unprecedented challenges for clustering algorithms if they are to be able to track changing clusters in noisy data streams using only the new data points because storing past data is not even an option. The harsh restrictions imposed by the *"you only get to see it once"* constraint on stream data calls for different computational models that may require us to borrow from the strength of different niches, such as robust statistical theory and statistical learning theory. An amalgam of such ideas can result in a framework for stream data mining that departs from traditional data mining. We proposed a new framework for learning synopses in evolving data streams. The synopses are based on hybrid learning strategies that combine the power of distributed learning with the robustness and speed of statistical and mathematical analysis. While the theoretical foundation of the proposed methods are generic enough to be applied in a variety of dynamic learning contexts, we focused on adaptive Web analytics and personalization systems that can handle massive amounts of data while being able to model and adapt to rapid changes, and while taking into account multi-faceted aspects of the data to enrich such applications. In particular, we proposed methods for: *(i)* Mining Evolving Data Streams, *(ii)* Mining Evolving Web Clickstreams, *(iii)* Integrating Multiple Information Channels, *(iv)* Higher-level Web Analytics, and *(v)* A Hybrid Meta-Recommender System.

Most *existing* Web mining and personalization methods *are limited to working at the level described to be the lowest and most primitive level in this chapter*, namely discovering models of the user profiles from the input data stream. We proposed to improve *understanding* of the *real intention behind clickstreams* by extending reasoning and discovery *beyond* the usual data stream level, and developing a new *multi-level framework* for Web usage mining and personalization that works at *several granularities, ranging from the user session or clickstreams, to the user profiles, to user profile communities, and finally user profile events and event sequences*, as depicted in Fig. 1. Our web mining and personalization framework also has the advantage of discovering multi-faceted user profiles that combine several information channels such as content, usage, and information need to enable a better understanding of the user's interests and to deliver more powerful hybrid recommendations.

Our preliminary work on variations of the components of the proposed mulitlayered web mining and personalization system has given promising results in mining evolving web clickstreams [27, 28], and on recommender systems [33, 34]. We plan to continue our investigation and validation efforts for the overall system.

References

1. P. Karnam A. Joshi, C. Punyapu. Personalization and asynchronicity to support mobile web access. In *Workshop on Web Information and Data Management, ACM 7th Intl. Conf. on Information and Knowledge Management*, Nov. 1998.
2. C. Aggarwal, J. Han, J. Wang, and P. Yu. A framework for clustering evolving data streams. 2003.
3. C. Aggarwal, J. Han, J. Wang, and P.S. Yu. A framework for clustering evolving data streams. In *Proc. 2003 Int. Conf. on Very Large Data Bases (VLDB'03)*, Berlin, Germany, Sept 2003.
4. S. Babu and J. Widom. Continuous queries over data streams. In *SIGMOD Record'01*, pp. 109–120, 2001.
5. M. Balabanovic and Y. Shoham. Fab: Content-based, collaborative recommendation. *Communications of the ACM*, 40(3):67–72, 1997.
6. D. Barbara. Requirements for clustering data streams. *ACM SIGKDD Explorations Newsletter*, 3(2):23–27, 2002.
7. J. Borges and M. Levene. Data mining of user navigation patterns. In H.A. Abbass, R.A. Sarker, and C.S. Newton, editors, *Web Usage Analysis and User Profiling, Lecture Notes in Computer Science*, pp. 92–111. Springer-Verlag, 1999.

8. P. Bradley, U. Fayyad, and C. Reina. Scaling clustering algorithms to large databases. In *Proceedings of the 4th international conf. on Knowledge Discovery and Data Mining (KDD98)*, 1998.

9. A. Buchner and M.D. Mulvenna. Discovering internet marketing intelligence through online analytical web usage mining. *SIGMOD Record*, 4(27), 1999.

10. R. Burke. Hybrid recommender systems: Survey and experiments. *User Modeling and User-Adapted Interaction*, 12(4):331–370, 2002.

11. R. Burke. Hybrid recommmender systems: Survey and experiments. *User Modeling and User-Adapted Interaction*, 12(4):331–370, 2002.

12. M. Charikar, L. O'Callaghan, and R. Panigrahy. Better streaming algorithms for clustering problems. In *Proc. of 35th ACM Symposium on Theory of Computing (STOC)*, 2003.

13. Y. Chen, G. Dong, J. Han, B.W. Wah, and J. Wang. Multi-dimensional regression analysis of time-series data streams. In *2002 Int. Conf. on Very Large Data Bases (VLDB'02)*, Hong Kong, China, 2002.

14. R. Cooley, B. Mobasher, and J. Srivastava. Web mining: Information and pattern discovery on the world wide web. In *IEEE Intl. Conf. Tools with AI*, pp. 558–567, Newport Beach, CA, 1997.

15. R. Cooley, B. Mobasher, and J. Srivastava. Data preparation for mining world wide web browsing patterns. *Journal of knowledge and information systems*, 1(1), 1999.

16. A.P. Dempster, N.M. Laird, and D.B. Rubin. Maximum likelihood from incomplete data via the em algorithm. *Journal of the Royal Statistical Society Series B*, 39(1):1–38, 1977.

17. U. Fayad, G. Piatetsky-Shapiro, P. Smyth, and R. Uthurusamy. *Advances in Knowledge Discovery and Data Mining*. AAAI/MIT Press, 1996.

18. S. Guha, N. Mishra, R. Motwani, and L. O'Callaghan. Clustering data streams. In *IEEE Symposium on Foundations of Computer Science (FOCS'00)*, Redondo Beach, CA, 2000.

19. G.H. Hardy, J.E. Littlewood, and G Pólya. *Inequalities*, chapter Tchebychef's Inequality, pp. 43–45. Cambridge University Press, Cambridge, England, 2nd edition, 1988.

20. M. Henzinger, P. Raghavan, and S. Rajagopalan. Computing on data streams, 1998.

21. P.J. Huber. *Robust Statistics*. John Wiley & Sons, New York, 1981.

22. H. Heckerman J. Breese and C. Kadie. Empirical analysis of predictive algorithms for collaborative filtering. In *14th Conf. Uncertainty in Artificial Intelligence*, pp. 43–52, 1998.

23. A. Joshi, S. Weerawarana, and E. Houstis. On disconnected browsing of distributed information. In *Seventh IEEE Intl. Workshop on Research Issues in Data Engineering (RIDE)*, pp. 101–108, 1997.

24. H. Mannila, H. Toivonen, and A.I. Verkamo. Discovering frequent episodes in sequences. In *Proceedings of KDD Congress*, pp. 210–215, Montreal, Quebec, Canada, 1995.

25. D. Mladenic. Text learning and related intelligent agents. *IEEE Expert*, Jul. 1999.

26. B. Mobasher, H. Dai, T. Luo, and M. Nakagawa. Effective personalizaton based on association rule discovery from web usage data. In *ACM Workshop on Web information and data management*, Atlanta, GA, Nov 2001.

27. O. Nasraoui, C. Cardona, C. Rojas, and F. Gonzalez. Mining evolving user profiles in noisy web clickstream data with a scalable immune system clustering algorithm. In *WebKDD 2003 KDD Workshop on Web mining as a Premise to Effective and Intelligent Web Applications*, Washington, DC, August 2003.

28. O. Nasraoui, C. Cardona, C. Rojas, and F. Gonzalez. Tecno-streams: Tracking evolving clusters in noisy data streams with a scalable immune system learning model. In *Third IEEE International Conference on Data Mining (ICDM'03)*, Melbourne, FL, November 2003.

29. O. Nasraoui and R. Krishnapuram. A new evolutionary approach to web usage and context sensitive associations mining. *International Journal on Computational Intelligence and Applications – Special Issue on Internet Intelligent Systems*, 2(3):339–348.

30. O. Nasraoui and R. Krishnapuram. One step evolutionary mining of context sensitive associations and web navigation patterns. In *SIAM conference on Data Mining*, pp. 531–547, Arlington, VA, 2002.

31. O. Nasraoui, R. Krishnapuram, H. Frigui, and Joshi A. Extracting web user profiles using relational competitive fuzzy clustering. *International Journal of Artificial Intelligence Tools*, 9(4):509–526, 2000.

32. O. Nasraoui, R. Krishnapuram, and A. Joshi. Mining web access logs using a relational clustering algorithm based on a robust estimator. In *8th International World Wide Web Conference*, pp. 40–41, Toronto, Canada, 1999.

33. O. Nasraoui and M. Pavuluri. Complete this puzzle: A connectionist approach to accurate web recommendations based on a committee of predictors. In *WebKDD- 2004 workshop on Web Mining and Web Usage Analysis , B. Mobasher, B. Liu, B. Masand, O. Nasraoui, Eds*, Seattle, WA, Aug 2004.

34. O. Nasraoui and C. Petenes. Combining web usage mining and fuzzy inference for website personalization. In *Proc. of WebKDD 2003 KDD Workshop on Web mining as a Premise to Effective and Intelligent Web Applications*, p. 37, Washington DC, August 2003.

35. M. Pazzani. A framework for collaborative, content-based and demographic filtering. *AI Review*, 13(5–6):393–408, 1999.

36. M. Perkowitz and O. Etzioni. Adaptive web sites: an ai challenge. In *Intl. Joint Conf. on AI*, 1997.

37. M. Perkowitz and O. Etzioni. Adaptive web sites: Automatically synthesizing web pp.. In *AAAI 98*, 1998.

38. R.O. Duda and P.E. Hart. *Pattern Classification and Scene Analysis*. John Wiley and Sons, 1973.

39. P.J. Rousseeuw and A.M. Leroy. *Robust Regression and Outlier Detection*. John Wiley & Sons, New York, 1987.

40. Robert E. Schapire. The boosting approach to machine learning: An overview. In *MSRI Workshop on Nonlinear Estimation and Classification*, 2002.

41. C. Shahabi, A.M. Zarkesh, J. Abidi, and V. Shah. Knowledge discovery from users web-page navigation. In *Proceedings of workshop on research issues in Data engineering*, Birmingham, England, 1997.

42. M. Spiliopoulou and L.C. Faulstich. Wum: A web utilization miner. In *Proceedings of EDBT workshop WebDB98*, Valencia, Spain, 1999.

43. J. Srivastava, R. Cooley, M. Deshpande, and P.N. Tan. Web usage mining: Discovery and applications of usage patterns from web data. *SIGKDD Explorations*, 1(2):1–12, Jan 2000.

44. L. Terveen, W. Hill, and B. Amento. Phoaks – a system for sharing recommendations. *Comm. ACM*, 40(3), 1997.

45. T. Yan, M. Jacobsen, H. Garcia-Molina, and U. Dayal. From user access patterns to dynamic hypertext linking. In *Proceedings of the 5th International World Wide Web conference*, Paris, France, 1996.

46. O. Zaiane and J. Han. Webml: Querying the world-wide web for resources and knowledge. In *Workshop on Web Information and Data Management, 7th Intl. Conf. on Information and Knowledge Management*, 1998.

47. O. Zaiane, M. Xin, and J. Han. Discovering web access patterns and trends by applying olap and data mining technology on web logs. In *Advances in Digital Libraries*, pp. 19–29, Santa Barbara, CA, 1998.

48. T. Zhang, R. Ramakrishnan, and M. Livny. Birch: An efficient data clustering method for very large databases. In *Proceedings of the ACM SIGMOD conference on Management of Data*, Montreal Canada, 1996.

Model Cloning: A Push to Reuse or a Disaster?

Maria Rigou[1,2], Spiros Sirmakessis[1], Giannis Tzimas[1,2]

[1] Research Academic Computer Technology Institute, 61 Riga Feraiou str.,
GR-26221 Patras, Hellas
[2] University of Patras, Computer Engineering & Informatics Dept., GR-26504,
Rio Patras, Hellas
{rigou,syrma,tzimas}@cti.gr

Abstract. The paper focuses on evaluating and refactoring the conceptual schemas of Web applications. The authors introduce the notion of model clones, as partial conceptual schemas that are repeated within a broader application model and the notion of model smells, as certain blocks in the Web applications model, that imply the possibility of refactoring. A methodology is illustrated for detecting and evaluating the existence of potential model clones, in order to identify problems in an application's conceptual schema by means of efficiency, consistency, usability and overall quality. The methodology can be deployed either in the process of designing an application or in the process of re-engineering it. Evaluation is performed according to a number of inspection steps. At first level the compositions used in the hypertext design are evaluated, followed by a second level evaluation concerning the manipulation and presentation of data to the user. Next, a number of metrics is defined to automate the detection and categorization of candidate model clones in order to facilitate potential model refactoring. Finally, the paper proposes a number of straightforward refactoring rules based on the categorization and discusses the aspects affecting the automation of the refactoring procedure.

1 Introduction

Modern web applications support a variety of sophisticated functionalities incorporating advanced business logic, one-to-one personalization features and multimodal content delivery (i.e. using a diversity of display devices). At the same time, their increasing complexity has led to serious problems of usability, reliability, performance, security and other qualities of service in an application's lifecycle.

The software community, in an attempt to cope with this problem and with the purpose of providing a basis for the application of improved technology, independently of specific software practices, has proposed a number of modeling methods and techniques that offer a higher level of abstraction to the process of designing and developing Web applications. Some of these methods

M. Rigou et al.: *Model Cloning: A Push to Reuse or a Disaster?*, Studies in Computational Intelligence (SCI) **14**, 37–55 (2006)
www.springerlink.com

derive from the area of hypermedia applications like the Relationship Management Methodology (RMM) [29], Araneus [2] and HDM [27], which pioneered the model-driven design of hypermedia applications. HDM influenced several subsequent proposals for Web modeling such as HDM-lite [21], a Web-specific version of HDM, Strudel [19] and OOHDM [45]. Extensions to the UML notation [9] to make it suitable for modeling Web applications have been proposed by Conallen [15, 16]. Finally, Web Modeling Language – WebML [10] provides a methodology and a notation language for specifying complex Web sites and applications at the conceptual level and along several dimensions.

Most of the above methodologies are based on the key principle of separating data management, site structure and page presentation and provide formal techniques and means for an effective and consistent development process, and a firm basis for re-engineering and maintenance of Web applications.

Deploying a methodology for the design and development of a Web application enhances effectiveness, but does not guarantee optimization in the design process, mainly due to the restricted number of available extreme designers/programmers [7]. Moreover, most applications are developed by large teams, leading to communication problems in the design/development process, often yielding products with large numbers of defects. In most of the cases due to the lack of time, designers reuse their previous work and experience without trying to fully adapt it to the requirements of the project at hand, resulting to "bad" cases of reuse. This situation stresses the need for restructuring/refactoring applications, even in their conceptual level, and the fact that effective modeling must be treated as a first class citizen and be considered from the very early and during all stages of the design process.

In the past, a number of research attempts have been conducted in the field of refactoring applications based on their design model. Most of them focus on standalone software artifacts and deploy UML to perform refactoring [39]. But despite the popularity of model-driven methodologies, there is an absence of assessment/analysis throughout the design and development process.

On the other hand, there are cases where the use of a methodology may cause a number of problems leading to design processes that make application development less effective and productive, particularly in the cases of automatic code generation. For this reason, there is a need for techniques appropriate for analyzing conceptual schemas so as to discover potential problems at the early phases of the design process and thus allow for fast recovery with the less possible effort. The goal of this paper is to argue the need to approach all aspects concerning effective design from the beginning in the Web application's development cycle. Since internal quality is a key issue for an application's success, it is important that it is dealt with through a design view, rather than only an implementation view.

2 Field Background and the Notion of Model Cloning

One of the intrinsic features of real-life software environments is their need to evolve. As the software is enhanced, modified, and adapted to new requirements, the code becomes more complex and drifts away from its original design, thereby lowering the quality of the software. Because of this, the major part of the software development cost is devoted to software maintenance [13, 28]. Improved software development methodologies and tools cannot resolve this problem because their advanced capabilities are mainly used for implementing new requirements within the same time slot, making software once again more complicated [40]. To cope with this increased complexity one needs techniques for reducing software complexity by incrementally improving the internal software quality.

Restructuring, refactoring and code cloning are well known notions in the software community. According to the reverse engineering taxonomy of [12], *restructuring* is defined as: "...the transformation from one representation form to another at the same relative abstraction level, while preserving the subject system's external behavior (functionality and semantics). A restructuring transformation is often one of appearance, such as altering code to improve its structure in the traditional sense of structured design. While restructuring creates new versions that implement or propose change to the subject system, it does not normally involve modifications because of new requirements. However, it may lead to better observations of the subject system that suggest changes that would improve aspects of the system."

In the case of object-oriented software development the definition of *refactoring* is basically the same: "...the process of changing a software system in such a way that it does not alter the external behavior of the code, yet improves its internal structure" [20]. The key idea here is to redistribute classes, variables and methods in order to facilitate future adaptations and extensions.

Code cloning or the act of copying code fragments and making minor, nonfunctional alterations, is a well known problem for evolving software systems leading to duplicated code fragments or *code clones*. Code cloning can be traced by *code smells* that is, certain structure in code that suggests the possibility of refactoring [20].

Roundtrip engineering has reached a level of maturity that software models and program code can be perceived as two different representations of the same artifact. With such an environment in mind, the concept of refactoring can be generalized to improving the structure of software instead of just its code representation.

There is a variety of techniques for supporting the process of detecting code cloning in software systems. Some of them are based on string and token matching [4, 18, 30, 31, 32, 37, 42], some others on comparing sub-trees and sub-graphs [6, 33, 34], while others are based on metrics characterization [1, 5, 33, 35, 38]. Moreover, there are a large number of tools that mine clones in source code and support a variety of programming languages such as C, C++, COBOL, Smalltalk, Java, and Python.

Clone mining in Web applications was first proposed by Di Lucca et al. [17] who study the detection of similar HTML pages by calculating the distance between page objects and their degree of similarity. Static page clones can also be detected with the techniques proposed by Boldyreff and Kewish [8] and Ricca and Tonella [41], with the purpose of transforming them to dynamic pages that retrieve their data form a database. Despite that, the decreasing percentage of Web sites that merely publish static content and the current shift towards Web applications with high degree of complexity, lead to the need to introduce new techniques, capable of coping with the problem. In the specific domain of Web application modeling, the notion of cloning has not yet been introduced. Even though a few model-level restructuring techniques have been proposed, this certain issue remains open for the scientific community [40]. Moreover, the existing techniques are based exclusively on UML as the modeling language and there is no technique based on one of the rest of Web application modeling languages and methods.

In this work we extend the notion of code cloning to the modeling level of a Web application. Analogously to code cloning, we introduce the notion of *model cloning* as the process of duplicating, and eventually modifying, a block of the existing application's model that implements a certain functionality. This ad-hoc form of reuse occurs frequently during the design process of a Web application. We will attempt to capture *model smells* that is, certain blocks in the Web application's model that imply the possibility of refactoring. Both notions were initially introduced in a previous work [44] addressing the specific domain of personalization in Web applications. The approach we follow in the current paper is much broader and refers to the generalized process of conceptual modeling.

More specifically, in this paper we provide a methodology in order to evaluate the conceptual schema of an application, by means of the design features incorporated in the application model. We try to capture cases (i.e. model clones) which have different design, but produce the same functionality, thus resulting in inconsistencies and ineffective design and may have been caused by inappropriate forms of model reuse.

The evaluation of the conceptual schema is performed in two steps of inspection: a first level evaluation of the hypertext compositions used in the hypertext design, and a second level evaluation of data manipulation and presentation to the user. We provide metrics for the quality evaluation of the application's conceptual schema and a number of restructuring/refactoring rules to improve the final applications quality.

Up to date, modeling of Web applications is mainly deployed during the early life cycle stages as a system analysis guide, with little attention allocated to the use of specifications at a conceptual level during application evaluation, maintenance and evolution analysis [25]. The proposed methodology can be deployed either in the process of designing an application or in the process of re-engineering it. In this work, WebML has been utilized as design platform for the methods proposed, mainly due to the fact that it supports a concrete framework for the formal definition of data intensive Web Applications and the fact that it is supported by a robust CASE tool called WebRatio [47]. Most of the work presented here can be generalized and applied to applications utilizing other modeling languages, such as OOHDM or HDM-lite, with slight straightforward modifications.

The remaining of this paper is structured as follows: Sect. 3 provides a short overview of WebML. Section 4 describes in detail the methodology for mining model clones in the conceptual schema of an application, while Sect. 5 introduces metrics for the evaluation of the clones. Finally, Sect. 6 provides refactoring suggestions and Sect. 7 concludes the paper and discusses future steps.

3 WebML: A Brief Overview

WebML is a conceptual model for Web application design [11]. It offers a set of visual primitives for defining conceptual schemas that represent the organization of the application contents and of the hypertext interface. These primitives are also provided with an XML-based textual representation, which allows specifying additional detailed properties that cannot be conveniently expressed in terms of visual notation. The organization of data is specified in WebML by exploiting the E-R model, which consists of entities (defined as containers of data elements) and relationships (defined as semantic connections between entities). WebML also allows designers to describe hypertexts for publishing and managing content, called *site views*. A site view is a specific hypertext, which can be browsed by a particular class of users. Within the scope of the same application, multiple site views can be defined.

Site views are internally organized into hypertext modules, called *areas*. Both site views and areas are composed of *pages*, which in turn include containers of elementary pieces of content, called *content units*. Typically, the data published by a content unit are retrieved from the database, whose schema is expressed by the E-R model. The binding between content units (and hypertext) and the data schema is represented by the *source entity* and the *selector* defined for each unit. The source entity specifies the type of objects published by the content unit, by referencing an entity of the E-R schema. The selector is a filter condition over the instances of the source entity, which determines the actual objects published by the unit. WebML offers predefined units, such as data, index, multidata, scroller, multichoice index, and hierarchical index (some of them are presented in Table 1), that express different ways of selecting entity instances and publishing them in a hypertext interface.

To compute its content, a unit may require the "cooperation" of other units, and the interaction of the user. Making two units interact requires connecting them with a *link*, represented as an oriented arc between a source

Table 1. Some basic WebML elements. The complete set is listed in [11]

Data unit	Multidata unit	Index unit	Hierarchical Index	Entry unit
Entity |Selector	Entity |Selector	Entity |Selector	Entity1 |Selector1 NEST Entity2 |Selector2	
Displays a set of attributes for a single entity instance.	Displays a set of instances for a given entity.	Displays a list of properties of a given set of entity instances.	Displays index entries organized in a multi-level tree.	Displays forms for collecting input data into fields

and a destination unit. The aim of a link is twofold: permitting navigation (possibly displaying a new page, if the destination unit is placed in a different page), and enabling the passing of parameters from the source to the destination unit.

Finally, WebML models the execution of arbitrary business actions, by means of operation units. An operation unit can be linked to other operation or content units. WebML incorporates some predefined operations (enabling content management) for creating, modifying and deleting the instances of entities and relationships, and allows developers to extend this set with their own operations.

4 A Methodology for Mining Model Clones

In what follows, we present a methodological approach for mining potential model clones at the conceptual level of a Web application. The methodology comprises three distinct phases.

- In the first phase, we transform the Web application's conceptual schema into a number of directed graphs, representing the navigation structure and the distribution of content among the areas and pages of the application. This forms the basis for an information extraction mechanism required for the next phase.
- In the second phase, we extract potential model clones and information related to the navigation and semantics of the application by utilizing graph mining techniques,
- Finally, in the third phase we provide a first level categorization of the potential model clones according to a number of criteria.

4.1 Conceptual Schema Transformation

In this phase the application's conceptual schema is preprocessed in order to provide the means for the potential model clones extraction. Assuming an application comprising a number of site views, we construct a first set of graphs representing the navigation, the content presentation and the manipulation mechanisms of the application. More specifically, we define a site view as a directed graph of the form $G(V, E, f_V, f_E)$, comprising a set of nodes V, a set of edges E, a node-labeling function $f_V : Vs \to \Sigma_V$, and an edge-labeling function $f_E : E \to \Sigma_E$. Function f_V assigns letters drawn from an alphabet Σ_V to the site view nodes, whereas f_E has the same role for links and the edge alphabet $\Sigma_E.\Sigma_V$ has a different letter for each different WebML element (content units, operations, pages, areas, etc). Correspondingly, Σ_E consists of all the different kinds of links (contextual, non contextual, transport & automatic). Besides the predefined WebML links, we introduce a special kind of edge (labeled "c") in order to represent the *containment* of content units or sub-pages in pages, as well as pages, sub-areas and operation units in areas. Note that there can be arbitrary containment sequences. A transformation example is depicted in Fig. 1 (Transformation A), where we transform a page containing several content units, interconnected by a number of contextual links.

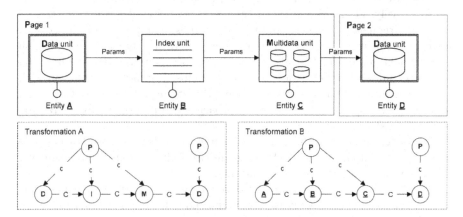

Fig. 1. Transformation of a WebML hypertext composition to its graph equivalents

Following a similar procedure, for every site view of the hypertext schema we create a second graph representing the data distribution within each area, sub-area and page, thus constructing a second set of graphs. In this case we define a site view as a directed graph of the form $Q(N, L, f_N, f_L)$, comprising a set of nodes N, a set of edges L, a node-labeling function $f_N : N \to \Sigma_N$, and an edge-labeling function $f_L : L \to \Sigma_L$. Function f_N assigns letters drawn from an alphabet Σ_N to the site view nodes, whereas f_L has the same role for links and the edge alphabet $\Sigma_L.\Sigma_N$ has a different letter for each different source entity used by the WebML elements comprising the hypertext schema, as well as for the pages, sub-pages, areas and sub-areas of the site view. Σ_L comprises all the different kinds of WebML links in order to model the context navigation within the hypertext schema. As in the previous transformation, we also introduce edges denoting containment. A transformation example is depicted in Fig. 1 (Transformation B).

4.2 Potential Model Clones Selection

Having modeled the navigation, content presentation and manipulation mechanisms of the application, as well as the data distribution within each site view, the next step is to capture model smells.

We traverse the first set of graphs constructed in the previous phase, in order to locate identical configurations of hypertext elements (subgraphs) along with their variants, either within a graph representing a single site view or among graphs representing different site views. The variants include all the alternatives in which the configuration retrieved starts and terminates (that is, all the nodes or sets of nodes in the graph passing the context and receiving context from the hypertext configuration). For every instance of the configuration we also retrieve the respective variants. The recovery of the various configurations can be achieved using graph mining algorithms.

Intuitively, after modeling the site views as directed graphs the task is to detect frequently occurring induced subgraphs. The problem in its general form is synopsized to finding whether the isomorphic image of a subgraph exists in a larger graph. The latter problem has proved to be NP-complete [26]. However, quite a few heuristics have been proposed to face this problem with the most prominent such approaches being the *gSpan* [48], the *CloseGraph* [49] and the *ADI* [46]. Any of the above approaches can be utilized for extracting the hypertext configurations.

Likewise, employing the same graph mining techniques, we traverse the second set of graphs in order to locate identical configurations of data elements (source entities) along with their variants.

Finally, we try to locate compositions of identical hypertext elements referring to exactly the same content interconnected with different link topologies. Ignoring the edges in the first set of graphs, except from those representing containment, we mine identical hypertext configurations within a graph or among graphs. Then, we filter the sets of subgraphs acquired utilizing the information represented in the second set of graphs (source entities), and keep those compositions that refer to the exact same data.

4.3 Potential Model Clones Categorization

In this phase, we categorize all the retrieved subgraph instances (potential model clones), in order to facilitate the quality evaluation procedure of the overall application conceptual schema, presented in the next section.

More precisely, for every instance of the hypertext configurations mined in the first case of graphs, we make a first level categorization according to the source entities and attributes that the WebML elements of the configurations refer to. To accomplish that, we utilize the information provided by the XML definition of each site view, where there is a detailed description of the source entities and the selectors of each element included in the site view [11]. For a specific configuration retrieved, we categorize its instances in the various site views of the application, as follows:

Configurations constituted by WebML elements referring to:

- exactly the same source entities and attributes,
- exactly the same source entities but different attributes (in the worst case the only common attribute is the object identifier (OID),
- partially identical source entities (i.e. Fig. 2),
- different source entities.

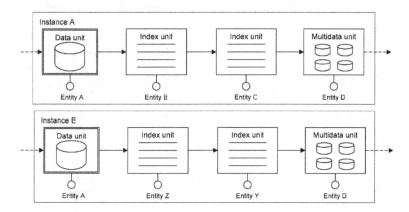

Fig. 2. Categorizing potential model clones at the Hypertext Level

We also categorize (exploiting the XML definition of site views) every instance of the data element configurations acquired by the graphs representing the data distribution as:

Configurations constituted by source entities utilized by:

- different WebML elements,
- similar WebML elements, that is elements of the same type such as composition or content management (e.g. in Fig. 3), entity A is utilized by two composition units, a multidata and an index).
- identical WebML elements.

The last category captures exactly the same hypertext configurations as the first case of the previous categorization.

Finally, potential model clones are the sets of hypertext configurations retrieved in the third step of the previous phase, where compositions of identical WebML elements referring to common data sources, utilizing different link topologies, have been identified (Fig. 4).

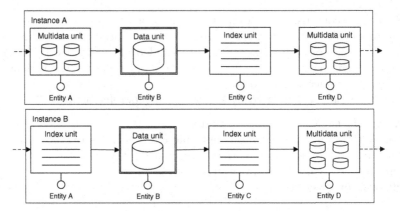

Fig. 3. Categorizing potential model clones at the Data Level

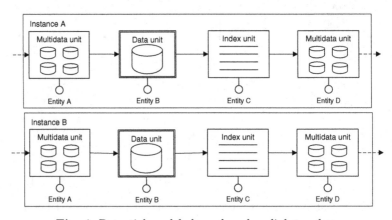

Fig. 4. Potential model clones based on link topology

5 Metrics and Quality Evaluation

Having identified the potential model clones at the hypertext and data representation level, we provide metrics for categorizing the application conceptual schema through a quality evaluation which specifies factors (referring to the identified clones) denoting possible need for refactoring. The evaluation is threefold: First, we evaluate the model clones by means of consistency, based on their variants throughout the conceptual schema. Second, we quantify the categorization presented in Sect. 4.3 based on the similarity among potential model clones and third we make an evaluation based on the topology of links.

5.1 Consistent Use of Model Clones

Fraternali et al. [24] have introduced a methodology for the evaluation of the consistent application of predefined WebML design patterns, within the conceptual schema of an application. We utilize this methodology and extend it, in order to introduce metrics depicting the consistent application of design solutions (i.e. model clones) retrieved during the first step of the procedure described in Sect. 4.2. To achieve that, we introduce two metrics that represent the statistical variance of the occurrence of N termination or starting variants of the identified model clones, normalized according to the best case variance. These metrics are called *Start-Point Metric (SPM)* and *End-Point Metric (EPM)* respectively. Following there is the formal definition of SPM (EPM is defined in an analogous way):

$$SPM = \sigma^2/\sigma_{BC}^2 \tag{1}$$

In (1) σ^2 is the statistical variance of the starting variants occurrences, which is calculated according to the following formula:

$$\sigma^2 = \left(\frac{1}{N}\right) \sum_{i=0}^{N} \left(p_i - \frac{1}{N}\right)^2 \tag{2}$$

where N is the number of variants addressed by the metric, while p_i is the percentage of occurrences for the *i-th* configuration variant. σ_{BC}^2 is the best case variance, and it is calculated by (2), assuming that only one variant has been coherently used throughout the application.

The last step in the metrics definition is the creation of a measurement scale, which defines a mapping between the numerical results obtained through the calculus method and a set of (predefined) meaningful and discrete values. According to the scale types defined in the measurement theory [43], the SPM metric adopts an ordinal nominal scale; each nominal value in the scale expresses a consistency level, corresponding to a range of numerical values of the metrics as defined in Table 2. The same scale covers the EPM metric as well.

5.2 Similarity of Model Clones

To evaluate the similarity between two hypertext configuration instances (potential model clones either with the same entities or the same WebML elements), we adopt the vector model and compute the degree of similarity [3].

Table 2. The SPM metric measurement scale

Metric's Range	Measurement Scale Value
0 <= SPM<0.2	Insufficient
0.2<= SPM<0.4	Weak
0.4<= SPM<0.6	Discrete
0.6<= SPM<0.8	Good
0.8<=SPM<=1	Optimum

To convert the initial configurations into vectors in the vector space model we define the vector $\vec{d}_i = (x_{i1}, x_{i2}, x_{i3} \ldots x_{in})$, where the compounds comprise all distinct WebML elements or entities of all configurations under consideration (for instance in Fig. 2, there are six distinct entities, while in Fig. 3, there are four). These compounds are considered as unigrams and are weighted by the frequency ϕ of each respective unigram. In the case of entities that appear with different attributes, their unigram is transformed to a fraction of frequency ϕ proportionate to the number of the participating attributes, as opposed to the number of all designed attributes. The vector space model proposes to evaluate the degree of similarity of a configuration d_m and a configuration d_μ (with the same entities or WebML elements) as the correlation between vectors $\vec{d}_m$ and $\vec{d}_\mu$. This correlation can be quantified by the cosine angle between these two vectors, that is:

$$\text{similarity}\,(d_m, d_\mu) = \cos(\vec{d}_m, \vec{d}_\mu) = \frac{\vec{d}_m \cdot \vec{d}_\mu}{\left|\vec{d}_m\right| \times \left|\vec{d}_\mu\right|} \Rightarrow$$

(3)

$$\text{similarity}\,(d_m, d_\mu) = \frac{\sum\limits_{\iota=1}^{t} w_{i,m} \times w_{i,\mu}}{\sqrt{\sum\limits_{\iota=1}^{t} w_{i,m}^2} \times \sqrt{\sum\limits_{\iota=1}^{t} w_{i,\mu}^2}}, \in [0,1]$$

where $\left|\vec{d}_m\right|$ and $\left|\vec{d}_\mu\right|$ are the norms of the two configurations.

Potential clones can be categorized as relevant or not, using the vector space model (that ranks them according to the degree of similarity between them). In the following, we propose the different thresholds and corresponding categorization based on the level of clones $similarity\,(d_m, d_\mu)$. This second stage of our evaluation approach uses as input the instances of the hypertext configurations retrieved. The goal of this stage is to check whether instances can be actually considered clones, and identify the opportunities of refactoring. Results of this step are depicted in ordinal scaled categories, as presented below.

Initially, configuration instances retrieved by the first set of graphs are classified according to the clone classification scheme shown in Table 3.

Classification follows an ordinal scale based on the degree of similarity between the potential clones. When the similarity degree increases, the probability having detected an actual model clone decreases. Therefore, the higher the level's value is, the smallest the probability to have identified a potential model clone gets.

Next, the configuration instances retrieved by the second set of graphs are classified according to the clone classification scheme as depicted in Table 4.

Likewise, the classification follows an ordinal scale based on the degree of similarity between the candidate model clones. Checking proceeds from level

Table 3. Classification of potential model clones based on the hypertext model

Level	Configurations of WebML Elements Referring to:
1	the same source entities and attributes.
2	source entities that are identical up to 75%.
3	source entities that are identical up to 50%.
4	source entities that are identical up to 25%.
5	totally different source entities.

Table 4. Classification of potential model clones based on the distribution of data

Level	Configurations Constituted by Source Entities Utilized by:
1	WebML elements of the same type, which are identical.
2	WebML elements of the same type, which are identical up to 75%.
3	WebML elements of the same type, which are identical up to 50%.
4	WebML elements of the same type, which are identical up to 25%.
5	totally different WebML elements.

1 to 5 and stops as soon as the adequate level has been recognized. Again, the probability to have identified a model clone increases as the level's value decreases.

5.3 Evaluation of the Link Topology

To further refine the evaluation procedure, we examine the hypertext configurations of candidate model clones within a page, based on the topology of their link connectivity. Intuitively, similarity of link topology at a local (within a page) or a broader (within an area or site view) hypertext level implies similarity in the business logic. When the differences are broader, then either the business logic is changed or potential misconfigurations are detected. We define *link topology similarity* as the cosine angle between vectors of the form $\overrightarrow{v}_i = (x_{i1}, x_{i2}, x_{i3} \ldots x_{in})$, where n is the number of different link connections in the hypertext model of candidate clones with the same WebML elements (in terms of instances and total number). The unigram compounds are assigned value "1" when the link is of the same direction, value "0" if the link does not exist at all, and value "−1" if the link creates an opposite direction element connection. As a result, we classify the hypertext configurations retrieved in the third step of the mining phase, in accordance to their link topology similarity as depicted in Table 5.

Table 5. Classification of potential model clones based on the links topology

Level	Description
1	Very high similarity in terms of link topology (equivalent).
2	75% similarity in terms of link topology (high).
3	50% similarity in terms of link topology (semi).
4	25% similarity in terms of link topology (low).
5	Totally different link topology (disjoint).

In this case also, we follow an ordinal scale for the classification scheme.

6 Refactoring Opportunities

After the classification of potential model clones according to their similarity, the hypertext architect has a first set of metrics pointing out "hot spots" for possible improvements in the conceptual schema of the application. The potential model clones are ranked according to their similarity level and their size (number of elements constituting the configuration[1]), thus forming a valuable tool providing quality improvement guidelines for the overall model. The methodology can be extended to support automatic model refinement, but there are a number of issues that need to be considered first. If an automated model refactoring process is to be deployed, checks for the syntactic and semantic correctness (e.g. conflicts, racing conditions, deadlocks) should be performed. Moreover, since personalization may be delivered in various forms and various levels of the design process (e.g. navigation or content personalization) automatic model refactoring could lead to inconsistencies with respect to initial functional requirements of the application. Nevertheless, a number of refactoring suggestions are straightforward.

Based on the first classification, the existence of two or more model clones belonging to Level 1 implies a very high refactoring opportunity. The designer should keep only one of the model clones, utilizing the "change siteview" unit if the clones are in different site views, or reconsider the link topology if the clones are in the same site view. In case that the WebML elements of two or more clones refer to exactly the same entities and differ to some of the attributes, merging all the attributes in every element should be considered (provided that content personalization is not affected). If the clones are categorized from levels 2 to 4, the designer should consider refactoring the hypertext schema by merging pages or areas.

On the other hand, the configurations categorized as level 5, might lead to capturing cases of effective application of a design solution or a design

---

[1] Larger configurations denote a higher probability of having identified a model smell.

pattern. These configurations imply effective use of previous experience that leads to a high degree of consistency within the application and promotion of usability. These are cases where the application gains in terms of quality. In this category of model clones, refactoring should be performed when SPM and/or EPM metrics values are under 0.6, by using where possible the variant having the larger occurrence frequency.

The various similarity levels computed during the second classification denote cases where the same data are presented to users (or manipulated by them) by means of different presentation (or content management) mechanisms. High similarity between clones implies "hot spots" where the designer should examine redesigning the hypertext in order to accomplish consistency in the presentation level and enhance usability of the final application.

Likewise, the last classification identifies cases in which the link topology of segments of the application model should be reconsidered, in order to accomplish consistency in the navigation and presentation level. Model clones belonging to the lower categorization levels should be considered for possibility of being merged by reordering links. Table 6 summarizes the refactoring proposals based on the categorization of model clone configurations that were presented in the previous sections.

7 Conclusions & Future Work

In this paper we have illustrated a methodology that aims at capturing potential problems caused by inappropriate reuse, within the conceptual schema of a Web application. We have introduced the notions of model cloning and model smells, and provided a technique for identifying model clones within an application's hypertext model by mapping the problem to the graph theory domain.

Moreover, we have specified a set of evaluation metrics quantifying the inappropriate reuse of clones and proposed refactoring recommendations. Applying the methodology can result in higher consistency and usability levels in the target application, as well as in a productivity increase in the application design and maintenance process.

Even though the quality of conceptual schemas depends at a high degree on the selected modeling language, the proposed methodology may be used by a series of languages with minor straightforward adjustments.

The proposed methodology extends the analytical application framework as presented by [14, 21, 22, 23, 24, 36], as apart from evaluating the consistency level in using design patterns, it also detects a number of factors that may cause design and quality problems at all levels of application modeling.

In the future we plan to apply the methodology to a large number of Web application conceptual schemas, in order to refine it and fine-tune the model clone evaluation metrics. We will also consider the distribution and effect of design patterns within a conceptual schema in accordance with the

Table 6. Refactoring proposals based on the similarity levels of the model clones

	Level	Description	Refactoring Proposals
Configurations of WebML elements referring to:	1	The same source entities and attributes.	Only one of the model clones should be kept, utilizing the "change siteview" unit if the clones are in different site views, or reconsider the link topology if the clones are in the same site view.
		Exactly the same entities with differences at some of the attributes.	Merging all the attributes in every element should be considered and deletion of the remaining clones.
	2	Source entities identical up to 75%.	Merging of pages and areas should be considered. Priority should be given to clones belonging to lower levels.
	3	Source entities identical up to 50%.	
	4	Source entities identical up to 25%.	
	5	Different source entities.	Consider refactoring when SPM and/or EPM metrics values are under 0.6, by using where possible the variant having the larger occurrence frequency.
Configurations constituted source entities utilized by:	1	WebML elements of the same type, which are identical.	Same as for level 1 in the previous case.
	2	WebML elements of the same type, up to 75% similar.	Examine redesigning the hypertext in order to accomplish consistency in the presentation level and enhance usability of the final application.
	3	WebML elements of the same type, up to 50% similar.	Priority should be given to clones belonging to lower levels.
	4	WebML elements of the same type, up to 25% similar.	
	5	Different WebML elements.	No refactoring should be considered.
Link Topology Similarity:	<2		Examine the possibility of merging clones by restructuring the link topology.

process of model clones identification. Moreover, we plan to investigate the representation of a conceptual model at a higher level of abstraction based on the existence of design patterns and model clones. Finally, we intend to extend the methodology in order to support automatic model refinement assuring at some level syntactic and semantic correctness, based on the evaluation results.

References

1. Antoniou G., Casazza G., Di Penta M., Merlo E. (2001) Modeling Clones Evolution Through Time Series. In the Proceedings of the International Conference on Software Maintenance, Florence, Italy, pp. 273–280
2. Atzeni P., Mecca G., Merialdo P. (1998) Design and Maintenance of Data-Intensive Web Sites. In the Proceedings of the 6th International Conference on Extending Database Technology – EDBT'98, pp. 436–450
3. Baeza-Yates R., Ribeiro-Neto B. (1999) Modern Information Retrieval. Addison Wesley, ACM Press
4. Baker B.S. (1995) On Finding Duplication and Near-Duplication in Large Software Systems. In the Proceedings of the Second Working Conference on Reverse Engineering, Toronto, Canada, pp. 86–95
5. Balazinska M., Merlo E., Dagenais M., Lague B., Kontogiannis K. (1999) Measuring Clone Based Reengineering Opportunities. In the Proceedings of the 6th IEEE International Symposium on Software Metrics, Boca Raton, USA, pp. 292–303
6. Baxter I.D., Yahin A., Moura L., Santa Anna M., Bier L. (1998) Clone Detection Using Abstract Syntax Trees. In the Proceedings of the International Conference on Software Maintenance, Washington DC, USA, pp. 368–377
7. Boehm B. (1981) Software Engineering Economics. Prentice Hall PTR
8. Boldyreff C., Kewish R. (2001) Reverse Engineering to Achieve Maintainable WWW Sites. In the Proceedings of the 8th Working Conference on Reverse Engineering (WCRE'01), Stuttgart, Germany, pp. 249–257
9. Booch G., Jacobson I., Rumbaugh J. (1998) The Unified Modeling Language User Guide. The Addison-Wesley Object Technology Series
10. Ceri S., Fraternali P., Bongio A. (2000) Web Modeling Language (WebML): a Modeling Language for Designing Web Sites. In the Proceedings of WWW9 Conference, Amsterdam
11. Ceri S., Fraternali P., Bongio A., Brambilla M., Comai S., Matera M. (2002) Designing Data-Intensive Web Applications. Morgan Kauffmann
12. Chikofsky E., Cross J. (1990) Reverse engineering and design recovery: A taxonomy. IEEE Software 7(1): 3–17
13. Coleman D.M., Ash D., Lowther B., Oman, P.W. (1994) Using Metrics to Evaluate Software System Maintainability. Computer 27(8): 44–49
14. Comai S., Matera M., Maurino A. (2002) A Model and an XSL Framework for Analysing the Quality of WebML Conceptual Schemas. In Proc of IWCMQ'02 – ER'02 International Workshop on Conceptual Modeling Quality, Tampere, Finland, October 2002
15. Conallen J. (1999) Modeling Web application architectures with UML. Communications of the ACM 42(10): 63–70

16. Conallen J. (1999) Building Web Applications with UML. Addison-Wesley, Reading MA

17. Di Lucca G.A., Di Penta M., Fasolino A.R., Granato P. (2001) Clone Analysis in the Web Era: an Approach to Identify Cloned Web Pages. In the Proceedings of the 7th IEEE Workshop on Empirical Studies of Software Maintenance, Florence, Italy, pp. 107–113

18. Ducasse S., Rieger M., Demeyer S. (1999) A language independent approach for detecting duplicated code. In the Proceedings of the International Conference on Software Maintenance, IEEE Computer Society Press, Oxford, UK, pp. 109–118

19. Fernandez M.F., Florescu D., Kang J., Levy A.Y., Suciu D. (1998) Catching the Boat with Strudel: Experiences with a Web-Site Management System. In the Proceedings of ACM-SIGMOD Conference, pp. 414–425

20. Fowler M., Beck K., Brant J., Opdyke W., Roberts, D. (1999) Refactoring: Improving the Design of Existing Code. The Addison-Wesley Object Technology Series

21. Fraternali P., Paolini, P. (1998) A Conceptual Model and a Tool Environment for Developing More Scalable, Dynamic, and Customizable Web Applications. In the Proceedings of the 6th International Conference on Extending Database Technology – EDBT'98, pp. 421–435

22. Fraternali P., Lanzi P.L., Matera M., Maurino A. (2004) Exploiting Conceptual Modeling for Web Application Quality Evaluation. In the Proceedings of WWW 2004 Alternate Tracks and Posters, New York, USA, May 2004

23. Fraternali P., Lanzi P.L., Matera M., Maurino A. (2004) Model-Driven Web Usage Analysis for the Evaluation of Web Application Quality. Submitted for publication to Journal of Web Engineering, April 2004

24. Fraternali P., Matera M., Maurino A. (2002) WQA: an XSL Framework for Analyzing the Quality of Web Applications. In the Proceedings of the 2nd International Workshop on Web-Oriented Software Technologies – IWWOST'02, Malaga, Spain, pp. 46–61

25. Fraternali P., Matera M., Maurino A. (2004) Conceptual-Level Log Analysis for the Evaluation of Web Application Quality. In the Proceedings of LA-Web Conference, IEEE Press, Cile, November 2004

26. Garey M.R., Johnson D.S. (1979) Computers and Intractability: A guide to NP-Completeness. New York: Freeman

27. Garzotto F., Paolini P., Schwabe D. (1993) HDM – A Model-Based Approach to Hypertext Application Design. ACM Transactions on Information Systems 11(1): 1–26

28. Guimaraes T. (1983) Managing Application Program Maintenance Expenditure. Communications of the ACM 26(10): 739–746

29. Isakowitz T., Sthor E.A., Balasubranian P. (1995) RMM: a methodology for structured hypermedia design. Communications of the ACM 38(8): 34–44

30. Johnson J.H. (1993) Identifying Redundancy in Source Code using Fingerprints. In the Proceedings of the CAS Conference, Toronto, Canada, pp. 171–183

31. Johnson J.H. (1994) Substring Matching for Clone Detection and Change Tracking. In the Proceedings of the International Conference on Software Maintenance, Victoria, Canada, pp. 120–126

32. Kamiya T., Kusumoto S., Inoue K. (2002) CCFinder: A Multilinguistic Token-Based Code Clone Detection System for Large Scale Source Code. IEEE Transactions On Software Engineering 28(7): 654–670

33. Kontogiannis K. (1997) Evaluation Experiments on the Detection of Programming Patterns Using Software Metrics. In the Proceedings of the 4th Working Conference on Reverse Engineering, Amsterdam, The Netherlands, pp. 44–54
34. Krinke J. (2001) Identifying Similar Code with Program Dependence Graphs. In the Proceedings of the 8th Working Conference on Reverse Engineering, Stuttgart, Germany, pp. 301–309
35. Lague B., Proulx D., Mayrand J., Merlo E.M., Hudepohl J. (1997) Assessing the Benefits of Incorporating Function Clone Detection in a Development Process. In the Proceedings of the International Conference on Software Maintenance, Bari, Italy, pp. 314–321
36. Lanzi P.L., Matera M., Maurino A. (2004) A Framework for Exploiting Conceptual Modeling in the Evaluation of Web Application Quality. In the Proceedings of the International Conference of Web Engineering-ICWE'04, LNCS 3140, Springer Verlag Publ, Munich, Germany, July 2004
37. Manber U. (1994) Finding Similar Files in a Large File System. In the Proceedings of the USENIX Winter 1994 Technical Conference, San Francisco, USA, pp. 1–10
38. Mayrand J., Leblanc C., Merlo E.M. (1996) Experiment on the Automatic Detection of Function Clones in a Software System Using Metrics. In the Proceedings of the International Conference on Software Maintenance, Monterey, USA, pp. 244–254
39. Mens T., Demeyer S., Du Bois B., Stenten H., Van Gorp P. (2003) Refactoring: Current research and future trends. In the Proceedings of the 3rd Workshop on Language Descriptions, Tools and Applications (LDTA 2003), April 6, Warsaw, Poland
40. Mens T, Tourwe, T. (2004) A survey of software refactoring. IEEE Transactions on Software Engineering, 30(2): 126–139
41. Ricca F., Tonella P. (2003) Using Clustering to Support the Migration from Static to Dynamic Web Pages. In the Proceedings of the 11th International Workshop on Program Comprehension, Portland, USA, pp. 207–216
42. Rieger M., Ducasse S. (1998) Visual Detection of Duplicated Code. In the Proceedings of the Workshop on Experiences in Object-Oriented Re-Engineering, Brussels, Belgium, pp. 75–76
43. Roberts D. (1999) Practical Analysis for Refactoring. Ph.D. thesis, University of Illinois at Urbana-Champaign
44. Sakkopoulos E., Sirmakessis S., Tsakalidis A., Tzimas G. (2005) A Methodology for Evaluating the Personalization Conceptual Schema of a Web Application. in the Proceedings of the 11th International Conference on Human-Computer Interaction (HCI International 2005), July 22–27, Las Vegas, Nevada, USA
45. Schwabe D., Rossi G. (1998) An object-oriented approach to web-based application design. Theory and Practice of Object Systems (TAPOS) 4(4): 207–225
46. Wang C., Wang W., Pei J., Zhu Y., Shi B. (2004) Scalable Mining of Large Disk-based Graph Databases. In the Proceedings of ACM KDD04, pp. 316–325
47. WebRatio (2005), available at: http://wwwwebratiocom
48. Yan X., Han J. (2002) gSpan: Graph-based substructure pattern mining. In the Proceedings of International Conference on Data Mining (ICDM'02), Maebashi, pp. 721–724
49. Yan X., Han J. (2003) CloseGraph: mining closed frequent graph patterns. In the Proceedings of ACM KDD03, pp. 286–295

Behavioral Patterns in Hypermedia Systems: A Short Study of E-commerce vs. E-learning Practices

A. Stefani, B. Vassiliadis, M. Xenos

Computer Science, Hellenic Open University, Greece
{stefani,bb,xenos}@eap.gr

1 Introduction

Web based systems are extremely popular to both end users and developers thanks to their ease of use and cost effectiveness respectively. Two of the most popular applications of web based systems nowadays are e-learning and e-commerce. Despite their differences, both types of applications are facing similar challenges: they rely on a "pull" model of information flow, they are hypermedia based, they use similar techniques for adaptation and they benefit from semantic technologies [3]. The underlying business models also share the same basic principle: users access digital resources from a distance without the physical presence of a teacher or a seller. The above mentioned similarities suggest that, at least, some user behavioral patterns are similar to both applications.

It is at least intriguing to notice that many published works relating to e-learning adaptation focus on machine learning techniques for log analysis or static user profiles with common characteristics. Despite the large number of related publications, very limited claims of success of such traditional adaptation techniques have been reported under real life e-learning situations [30]. Only within the past few years, researchers, originating mainly from the educational domain point out that little has been done for context-aware adaptation [20, 23, 28]. In the case of e-learning, context-aware parameters that seem to have been ignored to date, stem from behavioral and pedagogical theories. From our point of view, a simple but yet important parameter still eludes many e-learning adaptive hypermedia efforts: systems and techniques should be designed by both engineers and educationalists in order for research to bear fruits in real situations.

Although adaptability of web based systems has been the focus of a wide range of recent research efforts, these efforts are mainly focused on e-commerce. Most existing adaptation techniques for web-based systems are based on log analysis, user modeling or pre-determined rules [2]. E-learning

A. Stefani et al.: *Behavioral Patterns in Hypermedia Systems: A Short Study of E-commerce vs. E-learning Practices*, Studies in Computational Intelligence (SCI) **14**, 57–64 (2006)
www.springerlink.com   © Springer-Verlag Berlin Heidelberg 2006

adaptation has many common characteristics with e-commerce applications and many differences as well. This means that advances in e-commerce adaptation may be partially used for e-learning adaptation [8, 11, 13]. Adaptive Web-based e-learning systems are a class of adaptive Web systems, an attempt to replace the classic "one size fits all" approach to hypermedia.

In this work, we discuss some of the major similarities and differences between e-learning and e-commerce systems (and particularly B2C systems) that affect adaptation and system usability. First, a presentation of the state of the art in adaptation techniques will be presented. These techniques, focused more on the e-commerce field, are analyzed in the light of e-learning adaptation requirements. Special attention will be given to adaptation that relies on user modeling since a great deal of research effort has been placed on this area. We also suggest that strategies used for user modeling comprise the major difference between the two types of applications. Our methodology will then examine the different adaptation needs of e-learning and e-commerce in terms of diversity, context and background. Architectural issues are also of importance since different techniques impose different architectural styles for the same category of systems. Theoretical comparison suggests that crude e-commerce adaptation techniques are not very well suited for closed and formal e-learning systems. However, they are quite useful and cost-effective in open and informal systems. As formal systems we define applications that are used by organizations with well structured procedures such as Open or Virtual Universities. Informal learning systems are often free internet applications targeted to wide user populations. Finally, we conclude that machine learning techniques used for adaptation should take into account behavioral and educational theories for distance learning in order to have a serious impact.

2 Adaptive Hypermedia Systems: State of the Art

Adaptation in hypermedia systems followed initially two distinct approaches that later converged: adaptation based on rules and on algorithmic methods.

Rule based adaptation was dominating the field before algorithmic methods became a trend. A few years ago, most existing systems realized adaptation through the use of pre-determined rules, which assigned adaptation constituents to interaction situations in a rather "arbitrary" way. These rules are usually hard-coded in the user interface, and cannot be easily modified or reused across different applications. Additionally, there are no explicit representation of the goals underlying adaptation processes, and, in this sense, the latter cannot be taken explicitly into account in the adaptation process. Decision-theoretic frameworks for run-time adaptation, are mostly utility-based decision making techniques in the context of the standard reference model for intelligent hypermedia presentation systems. Recent trends examine the efficiency of hybrid rule-based/algorithmic methods for adaptation [15].

A very popular, algorithmic approach for discovering user behavior in hypermedia systems is a special area of data mining: web mining [12]. In turn, Web mining is generally focused on content, structure and usage mining [4]. Web mining relies on standard data mining algorithmic approaches which face two major problems in this particular application domain: weak relations between users and the hypermedia system, and complicated behaviors [27]. The first problem is created by the fact that users often do not have clear goals or the system does not have enough information about them to adapt to their preferences (as in the case of first time or infrequent users). The second one is caused by the diversity of the user population.

Content mining analyses the content of hypermedia documents using information retrieval techniques. Usage mining initiated largely by approaches like the one of [22], gained greater attention and presented significant results in the next few years [10, 17, 21]. A special area of web usage mining is collaborative (social) filtering [13] which refers to the categorization users to groups according to their preferences. The most popular methods are Markov models, association rules, sequential patterns, most forward access patterns [5], tree structure access patterns [31], clustering and hybrid models.

User modelling is a popular adaptation method and already counts several years of application to a variety of systems. A user model maintains an explicit and dynamic representation of the user. It represents the system's understanding of a user and it consists both of known facts about the user (such as personal information) and inferred beliefs based on previous interactions. A typical model classifies user-related information into four main categories [26]:

- Personal information (e.g. age, sex, preferences)
- Information about how the user interacted with the offered services (e.g. path used)
- Information about services the user has used
- Explanation of the result of specific service actions (e.g. unsuccessful buying attempts)

E-learning was between the first category of applications, and probably the first among the on-line ones, that used user models for adaptation. Early approaches used machine learning techniques initially developed for the information system domain to support adaptation. The advent of the internet and the spectacular increase of the user population, both in number and diversity, posed the need for new approaches. New methods focus on the way in which learners acquire, store, process and share knowledge rather than forcing them to follow a stereotype, predetermined behavior [19, 25].

Learner (student) modeling has been extensively researched because there is already a strong background on pedagogical theories [6, 7]. The most interesting approaches include the overlay, the perturbation, the analytic, the synthetic and the mixed model.

Static buyer profiles were the primary source of knowledge in early implementations of adaptive B2C applications. A striking difference with student modeling is that comparatively little progress has been made in the development of user modeling components for e-commerce systems [24].

Besides traditional approaches that produce user models based on log file analysis or questionnaires, works worth mentioning are those applied to live help systems for e-commerce web sites (like the one of [1]) and recommender mechanisms.

Although we will not discuss in detail these approaches, it is obvious that there is no underlying theory in constructing buyer models nor any standardization efforts. Although machine learning does offer solutions to some problems of buyer modeling, it cannot be considered as a panacea with the argument that there are no other approaches/theories to be used.

3 Adaptation Requirements: E-learning vs. E-commerce

A fundamental question in this comparison study is 'what should be adaptable and how?' [18]. In this section, we explore requirements that affect the design of adaptation mechanisms.

The answer to the question of what can be adapted is mainly of a technological nature, while the one referring to how, a pedagogical – technological one. Content, layout and navigation are the three types of adaptation common in most adaptive hypermedia systems. Adaptation strategies should also be context specific. In the case of e-commerce, they rely on behavioral, social, marketing and other theories. In the case of e-learning, they rely on pedagogical and learning theories.

User behavior differs significantly. According to [16], B2C user behavior falls under four general categories: directed buying, search/deliberation, hedonic browsing and knowledge-building. The first two are goal-oriented, while the other two resemble explorative search behavior. On-line learning behavior, on the other hand, may follow the social/collaborative, contextualized or experiential model. There is an obvious difference in complexity but also an interesting point: knowledge building is also used in e-commerce: navigation and hyperlinked data aid collaborative knowledge building which is becoming available for all kinds of adaptive hypermedia systems [29].

It is obvious that, although both e-learning and e-commerce are largely hypermedia based and use the same internet protocols and technologies to work, user goals differ significantly. Starting from e-learning applications, user goals are, ideally, to reach a set of predefined educational objectives, to learn. These objectives are the same for both the designers and the users of such system. E-learning is closely connected to educational, pedagogical and behavioral theories. Most current applications follow the information transfer paradigm where information is passed from the system to the user. Advanced learning models, the future trends in e-learning, anticipate knowledge construction and sharing and most importantly collaboration through the formation and management of virtual learning communities. Most formal platforms support the on-line presence of tutors (either real persons or avatars). Since e-learning uses on-line resources in its core mechanism, requirements such as efficient browsing and searching mechanisms are necessary. But how can context be described without the proper metadata? Semantic enriched, context-aware hypermedia may be the solution.

In the case of e-commerce and especially B2C (Business to Consumer) applications, goals are somewhat different for designers and users. Designers follow marketing strategies in order to sell as many products as possible. This includes making the user experience as seamless as possible, recommender systems and automatic offers. User goals are the same as every buyer goals: locate the appropriate product as simply as possible and access as much and relevant information as possible. It is obvious that designers and users, in this context, do not share the same goals in so far as an underlying theory is concerned. Marketing theories are followed by designers for increasing profits, not by users to directly gain benefit. In B2C, information is transferred from the application to the user and there is no direct possibility of knowledge building as in the case of e-learning. There is, however, the support of knowledge sharing through the use of off-line collaboration (mainly forums where buyers exchange opinions). Another difference is the absence of on-line guidance in the strict and formal form of tutoring. Off-line consultation with experts, or on-line support is significant but not as crucial as in the case of e-learning. On-line communities are present in the form of communities of practice and share significant information. They are however, in most cases, informal and communicate mostly by off-line tools.

New trends in pedagogy concentrate more on constructivism (Duffy and Jonassen, 1992), the building of knowledge by way of social interaction and collaboration on-line. Although constructivism has been identified by many researchers as one of the most appropriate methods for learning, science has not yet comprehended and analyzed the mental processes of human knowledge building, collaboration, sharing, evolution and reuse. Thus, a learner does not behave exactly as a buyer does. Another important parameter is the identification of how users perceive and process information and how they complete tasks. It seems that the "one-size fits all" approach has proved to be relatively successful in e-commerce, which is not the case in e-learning. The number of different learning styles is large, and each of them is largely affected by numerous context parameters known only to the teaching staff. So, traditional adaptation mechanisms are somewhat superficial. One interesting proposal by [23] is to let the teaching staff configure adaptation by choosing the appropriate learning style that best describes the educational context. This approach has merit in theory, but it is difficult to be accomplished in practice since such software may become too difficult to use. Flexibility is the key to the success of this approach.

A popular misconception is that adaptation in both e-learning and e-commerce is governed by the same principles [14]. Techniques such as log analysis with the use of machine learning techniques, or general purpose user profiling will do the job in both cases. Although the above mentioned strategies are quire successful in the e-commerce paradigm, this is not always the case in e-learning. Most efforts fail to take under consideration pedagogical models and educational goals. E-learning adaptation in formal systems is more about sequencing learning material and workflow of learning processes. E-commerce recommendation/adaptation mechanisms simply will not do because they rely on common beliefs (preferences) which are often misconceptions and possibly not quite useful pedagogically. For example, the logical structure of a course is not taken into account when links or documents are recommended by an adaptation mechanism. This way, a pedagogically false sequence may be initiated by common user mistakes. E-learning is a procedure that is guided by formal theories, with well defined goals and methods in order to impose some kind of pedagogical control over the learning process.

These are some of the differences that drive the diversity of user behavior in e-learning and e-commerce.

4 Conclusions

User behaviour is diverse in e-learning and e-commerce hypermedia applications. Furthermore, different research approaches have flourished in these domains: in the former user modeling and in the latter machine learning. The main problem in current implementations is that these techniques are used in a straightforward way without any tailoring. E-learning adaptation uses machine learning techniques mostly used in e-commerce resulting in poor efficiency. E-commerce has benefited from student modeling approaches but missing underlying theories produce static user models.

In this work, we reviewed the state of the art in discovering user behavior in both hypermedia contexts and briefly discussed differences and similarities. We argued that data mining techniques used in e-commerce should be used as a basis for e-learning. In fact, they should be combined with pedagogical approaches and theories.

Although the field of hypermedia adaptation is huge we believe that we made a small contribution to the endeavor of real adaptation in hypermedia. Our main argument, although simple but overlooked, is that adaptation should be context-specific and thus be in accordance with the underlying theory. Future work includes a more thorough investigation of social, cultural and economic factors that impose diversity in adaptive hypermedia systems and some recommendations particularly for e-learning and e-commerce applications. Furthermore, to investigate how new architectural models such as service oriented computing affect traditional adaptation in hypermedia.

Acknowledgements

This work was supported by the EPEAEKII Iraklitos programme (code: 88738) at HOU.

References

1. Aberg J., Shahmehri N., Maciuszek D. (2001). User modelling for live help systems: initial results. Proceedings of the 3rd ACM conference on Electronic Commerce, pp. 194–197
2. Brusilovsky P., Maybury M.T. (2002). From adaptive hypermedia to adaptive Web. Communications of the ACM, 45(5): 31–33
3. Brusilovsky P., Nejdl W. (2004). Adaptive hypermedia and adaptive web. In:Singh M., (ed) Practical Handbook of Internet Computing. Chapman & Hall/CRC
4. Chakrabarti S., (2002). Mining the Web: Analysis of Hypertext and Semi Structured Data. Morgan Kaufmann
5. Chen MS., Jong J.S., Yu P.S. (1998). IEEE Transactions on Knowledge and Data Engineering, 10(2):209–221
6. De Bra P. (2002). Adaptive educational hypermedia on the web. Communications of the ACM, 45(5):60–61
7. De Bra P., Brusilovsky P., Houben G.J. (1999). Adaptive hypermedia: from systems to framework. ACM Computing Surveys (CSUR), 31(4), Article No. 12
8. Dolog P., Henze N., Nejdl W., Sintek M., (2004). Personalization in Distributed e-Learning Environments. In Proceedings of 13th World Wide Web Conference, pp. 170–179
9. Duffy T.M., Jonassen D.H. (1992). Constructivism and the Technology of Instruction: A Conversation. Lawrence Erlbaum Associates.
10. Eirinaki M., Vazirgiannis M. (2003). Web mining for web personalization. ACM Transactions on Internet Technology (TOIT), 3(1):1–27
11. Gams E., Reich S. (2004). An analysis of the applicability of user trails in web applications. 2004 Web Engineering Workshop
12. Han J., Kamber M. (2001). Data Mining: Concepts and Techniques. Morgan Kaufmann, San Francisco
13. Kim D.H., Atluri V., Bieber M., Adam N., Yesha Y., (2004). Web personalization: A clickstream-based collaborative filtering personalization model: towards a better performance. Proceedings of the 6th annual ACM international workshop on Web information and data management, pp. 88–95
14. Li L., Zaiane O.R. (2004) Combining Usage, Content and Structure Data to Improve Web Site Recommendation. 5th International Conference on Electronic Commerce and Web Technologies (EC-Web 04), Springer Verlag LNCS 3182, pp. 305–315
15. Mendes E., Mosley N., Counsell S. (2003). Do adaptation rules improve web cost estimation?. Proceedings of the fourteenth ACM conference on Hypertext and hypermedia, Nottingham, UK, pp. 173–183
16. Moe W. (2001) Buying, searching, or browsing: Differentiating between online shoppers using in-store navigational clickstream. Journal of Consumer Psychology, 13(1&2):29–40

17. Pierrakos D., Paliouras G., Papatheodorou Ch, Spyropoulos C.D. (2003). Web Usage Mining as a Tool for Personalization: A Survey. User Modeling and User-Adapted Interaction, 13(4):311–372

18. Rumetshofer H., Wöß W. (2003). An approach for adaptable learning systems with respect to psychological aspects. 2003 ACM Symposium on Applied Computing (SAC 2003), pp. 558–563

19. Schewe K.D., Thalheim B., Binemann-Zdanowicz A., Kaschek R., Kuss T., Tschiedel B (2005). A Conceptual View of Web-Based E-Learning Systems. Education and Information Technologies, 10(1&2):83–110

20. Spiliopoulou M., Pohle C., Teltzrow M. (2002). Modelling Web Site Usage with Sequences of Goal-Oriented Tasks. Multikonferenz Wirtschaftsinformatik, in: E-Commerce – Netze, Märkte, Technologien, Physica-Verlag, Heidelberg

21. Srikant R., Yang Y. (2001). Mining Web Logs to Improve Web Site Organization. Proc. Of the WWW01 Conference, pp. 430–437

22. Srivastava J., Cooley R., Deshpande M., Tan P.N. (2000), Web usage mining: discovery and applications of usage patterns from Web data. ACM SIGKDD Explorations Newsletter, 1(2):12–23

23. Stash N., Cristea A., De Bra P. (2004). Authoring of Learning Styles in Adaptive Hypermedia: Problems and Solutions. WWW 2004, May 17–22, 2004, New York, New York, USA, pp. 114, 123

24. Strachan L., Anderson J., Sneesby M., Evans M. (2000). Minimalist User Modelling in a Complex Commercial Software System. User Modeling and User-Adapted Interaction, 10(2-3):109–146

25. Tsiriga V., Virvou M. (2004). A Framework for the Initialization of Student Models in Web-based Intelligent Tutoring Systems. User Modeling and User-Adapted Interaction, 14(4):289–316

26. Vassiliadis B., Makris C., Tsakalidis A., Bogonikolos N. (2003). User Modelling for Adapting and Adaptable Information Retrieval. Journal of Applied System Studies, 4(1)

27. Wang L., Meinel C. (2004). Behaviour Recovery and Complicated Pattern Definition in Web Usage Mining. In: Koch N., Fraternali P., Wirsing M (eds.) LNCS 3140. Springer-Verlag Berlin Heidelberg, pp. 531–543

28. Wolf C. (2003). iWeaver: Towards 'Learning Style'-based e-Learning in Computer Science Education. Australasian Computing Education Conference (ACE2003), Adelaide, Proceedings of the fifth Australasian computing education conference on Computing education 2003 – Volume 20, pp. 273–279

29. Wu H., Gordon M.D., DeMaagd K., Bos N. (2003). Link analysis for collaborative knowledge building. Proceedings of the fourteenth ACM conference on Hypertext and hypermedia, Nottingham, UK, pp. 216–217

30. Xenos M., Pierrakeas C, Pintelas P. (2002). Survey on Student Dropout Rates and Dropout Causes Concerning the Students in the Course of Informatics of the Hellenic Open University. Computers & Education, 39(4):361–377

31. Zaki M.J. (2002). Efficiently Mining Frequent Trees in a Forest. Proceedings of the eighth ACM SIGKDD international conference on Knowledge discovery and data mining, Edmonton, Alberta, Canada, pp. 71–80

Adaptive Personal Information Environment Based on Semantic Web

Thanyalak Maneewatthana, Gary Wills, Wendy Hall

Intelligence, Agents, Multimedia Group, School of Electronics and Computer Science, University of Southampton, SO17 1BJ, United Kingdom
{tm03r, gbw, wh}@ecs.soton.ac.uk

Abstract. In order to support knowledge workers during their tasks of searching, locating and manipulating information, a system that provides information suitable for a particular user's needs, and that is also able to facilitate the sharing and reuse information is essential. This paper presents Adaptive Personal Information Environment (a-PIE); a service-oriented framework using Open Hypermedia and Semantic Web technologies to provide an adaptive web-based system. a-PIE models the information structures (data and links), context and behaviour as Fundamental Open Hypermedia Model (FOHM) structures which are manipulated by using the Auld Linky contextual link service. a-PIE provides an information environment that enables users to search an information space based on ontologically defined domain concepts. The users can add and manipulate (delete, comment, etc) interesting data or parts of information structures into their information space, leaving the original published data or information structures unchanged. a-PIE facilitates the shareability and reusability of knowledge according to users' requirements.

1 Introduction

Knowledge management and the associated tools aim to provide an environment in which people may create, learn, share, use and reuse knowledge, for the benefit of the organisation, the people who work in it, and the organisation's customers. However, instead of helping users, many systems are just increasing the information overload. Adapting text and sets of relationships of information or contents to the needs of individual users greatly enhances navigation and comprehension of information spaces.

The Semantic Web can be used to organise information in concept structures, while web services allow the encapsulation of heterogeneous knowledge and modularization of the architecture. In addition, web services also support dynamic and shareable frameworks for automated adaptation [1].

An ontology can be used to enrich the semantics of data and information structures to aid the process of information searching (use and reuse of

T. Maneewatthana et al.: *Adaptive Personal Information Environment Based on Semantic Web*, Studies in Computational Intelligence (SCI) **14**, 65–74 (2006)
www.springerlink.com

information). In this paper we propose an Adaptive Personal Information Environment system (a-PIE), a service-oriented framework for reusability and shareability of information. a-PIE aims to provide a system in which members of the community or organisation are able to browse information tailored to their needs, store relationships to content of interest in their own information repository, and are able to augment the relationships for reuse.

The background of related technologies; Open Hypermedia, Fundamental Open Hypermedia Model, Auld Linky, Semantic Web, and web services are briefly described in the next section. Then, a system overview of a-PIE is presented, focusing on the support for adaptation, reusability and shareability of information. Finally some conclusions and future work are presented.

2 Background

In Open Hypermedia Systems, links are considered as first-class objects. These entities are manipulated separately from hypermedia documents and stored independently in link databases (linkbases). Links and data are then added to hypermedia documents by means of a link service. The advantages of the link service approach are that links can be created, added, and edited without affecting the original document. By moving hyperlinks out of documents and into link databases, the relationships between documents are separated from the document content [4]. Therefore the collection of documents becomes more maintainable, quicker to produce and easier to reuse. Changes in target documents only require changes in linkbases. So using external linkbases enables different sets of links over the same content for different audiences and tasks. In addition, individuals and groups can maintain their own personal link databases.

The Fundamental Open Hypermedia Model (FOHM) [2] is a protocol for open hypermedia with additional context-awareness features. It is a data model for expressing hyperstructure by representing associations between data. Auld Linky is a context based link server which supplies links from specified linkbases by parsing FOHM structures. The four essential components of a FOHM structure are:

- *Data objects*: Data objects are wrappers for any piece of data that lies outside of the scope of the FOHM model.
- *Associations*: Associations are structures that represent relationships between Data objects.
- *References*: Reference objects are used to point at Data objects or at parts of Data objects.
- *Bindings*: Bindings specify the attributes of the connection between Association and Data objects.

FOHM also provides two modifier objects, which can be attached to any part of the FOHM structure, these are Behaviour and Context. Behaviour objects are used by client applications, whereas Context objects define conditions for the visibility of particular objects. Context objects also define the description type and are used by Auld Linky to distinguish which bindings should be returned to the user.

Web Services are software systems which provide standard ways to interoperate between various existing applications run on heterogeneous resources or frameworks. Web Services have been designed to wrap existing applications and expose them using an interface described in machine-processable format: Web Services Description Language (WSDL). Other systems can interact with web services using Simple Object Access Protocol (SOAP) messages. The use of Web Services can be described as loosely coupled, reusable software components, which can be orchestrated on the fly [3].

The key idea of the Semantic Web is to have data defined and linked in such a way that its meaning is explicitly interpretable by software processes rather than just being implicitly interpretable by humans [4]. The Semantic Web can represent knowledge, including defining ontologies as metadata of resources. An ontology is a means to describe formally a shared understanding, and capture knowledge for a particular domain [5]. It will be necessary to annotate web resources with metadata to provide some indication of its content.

Community portals are information portals designed to support and facilitate a community of interest. They allow members of a community to contribute information either by submitting or posting the information to the system.

3 System Overview

The adaptive Personal Information Environment (a-PIE) aims to provide a system in which members of the community are able to browse information suitable to their particular needs, identify and store FOHM structures in their own information repository which users may enhance prior to reuse. Thereby enable the sharing and reuse of structures and data. a-PIE further enhances these functionalities by using ontologies to define the associations and facilitate interoperate between knowledge components.

a-PIE consists of several services. The domain concept service provides the relevant concept. The user model service updates user model. The data item and association service manipulates data and structures (associations), as FOHM objects, from linkbases through the contextual link server (Auld Linky). The user service or adaptive engine provides the facilities for reconciling the data content, FOHM structures, and user model, to present the individualised document to the user through a web browser.

Figure 1 illustrates the system architecture of a-PIE. The functionality of the system is made available to software agents through a Web Service interface (WSDL), and to end-users through a Web browser. The input into

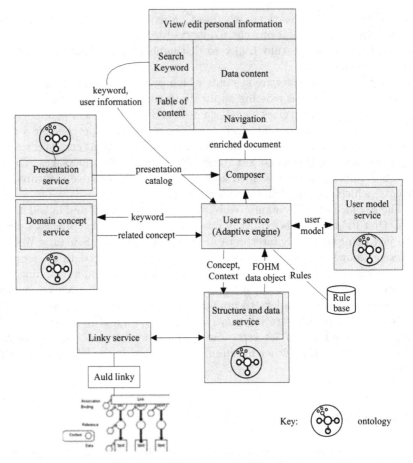

Fig. 1. a-PIE System Architecture

the system is a collection of data objects from a user and the output is an enriched document customised to a particular user's needs.

The following section presents, how a-PIE supports data creation, storage, adaptation, reusability and sharing of information.

3.1 Data Creation, Storage and Adaptation

In order to promote reusability and sharing of information, the system separates these components; domain concept, data, structure, presentation, user information, context and behaviour. Each model has an ontology as a means for interoperation.

- *Domain concept*: represents the basic structure of concept schemes. Simple Knowledge Organisation System (SKOS) [6] is used to express the basic structure of concept.

- *Domain data model*: represents the data created and stored by individual or organisations on their own web sites, in the form of FOHM Data objects.
- *Structure model*: connects FOHM Data objects defined in XML format as a series of FOHM Association structures. FOHM Association structures can be used to model several hypermedia structures; Navigational Link, Tour, Level of Detail, and Concept.
- *User model:* represents user-related information, such as background knowledge, preference, and information about user.
- *Context model*: represented by FOHM Context objects which can be attached to a Data or Association object for describing the context in which the data item or association is visible (or hidden) from the user.
- *Behaviour model*: represented by FOHM behaviour objects which can be attached to a Data or Association object. Behaviour objects describe an action that occurs as a result of an event.
- *Presentation model*: display and machine-related information, such as the colour schemes for resource presentation.

Structures that can be represented by FOHM are:

- *Navigational link*: The navigational link is an association with typed data items, source, destination or bi-directional locations.
- *Tour*: The tour is an association that represents an ordered set of objects.
- *Level of Detail*: The Level of Detail structure represents an ordered sequence of objects, where each object represents similar conceptual information with increasing of complexity.
- *Concept:* The concept is a collection of objects, such as text, image, or audio, which represent the similar conceptual information.

a-PIE provides adaptive hypermedia support through the use of Auld Linky [2] as a contextual link server to integrate FOHM structure and data objects according to the context. From Brusilovsky's taxonomy [7] of adaptive hypermedia techniques; adaptive navigation and presentation, FOHM structures can be combined to implement a range of these techniques [8]. An example of FOHM structure is shown in Fig. 2. This is a link with one source (the location with the word "FOHM") and two destinations (with urls). Both destinations explain "FOHM", the first with none technical information while the second with technical detail. If the structure was loaded into Auld Linky and queried using this context then Auld Linky would remove the inappropriate destination.

Therefore, this system can produce the information suitable to users' needs as an adaptive web-based system. Once data is made available by the organisation and published through the web site, other users can use, reuse, or browse. This means that anyone with an Internet connection can use and reuse the information.

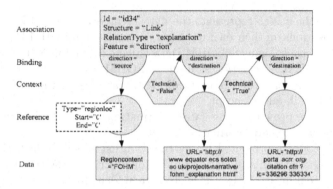

Fig. 2. A simple FOHM Navigation Link

3.2 Sharing, Reusing and Enriching the Information

In Open Hypermedia links are first-class objects and manipulated independently. Open Hypermedia makes links (associations) between different pieces
of information (e.g., images or text). In a-PIE, the FOHM models are used as
a protocol from open hypermedia. Hypermedia link-types describe the associations. Therefore, hypermedia link-types are knowledge relations [9], and a
set of hypermedia link types may be represented as an ontology. In a-PIE, an
ontology of link types has created based on relationships suggested by Bieber
and Yoo [10].

Sharing and reuse of information are integral aspects of the Semantic Web.
In a-PIE, the ontology is based on Semantic Web technology standards (RDF
[11]/OWL [12]) and is the backbone of the system. The ontologies represent
relationships of domain concepts. The ontologies are also used to enrich links
and data content, and to enable other users or organisations to share and reuse
the content or structure of the FOHM representation. In particular, each user
in a community or organisation can browse and search the site. In addition,
the user can; add data, add additional specialist information, and change the
context or behaviour of the FOHM structures. The users might also use their
own domain concept, context or behaviour for categorising or describing the
information.

Figure 3 illustrates a simple scenario on how information can be used and
reused in a-PIE. The browser is divided, apart from menu and navigation
areas, into three main areas; search, table of contents and data content areas.
The data content area displays the data content stored as FOHM-data objects. In this scenario, a user initiates a query, after processing the results are
returned to the user according to their profile (the level of detail and types of
user). The users add notes or comments to a particular piece of information.

The processes for browsing, adding structures the user is interested in and
adding more data content (such as notes or comments) to a particular item
are described in the following sequence of operations.

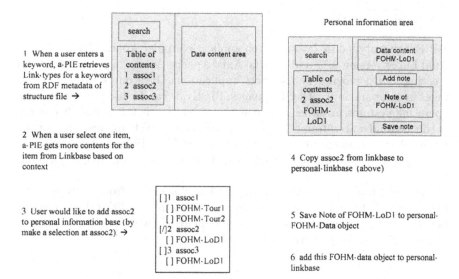

Fig. 3. Simple processes for a adding note to particular data item

1. A user enters a keyword into a search box. Then, the system will find the associations that have relationships related to this concept from metadata of FOHM associations stored in a RDF file, and then show the results as table of contents.
2. When the user selects the association they are interested in, the system will get more *structure of contents* from the linkbase, represented as FOHM associations.
3. The data content stored in FOHM data object will be displayed in the data content area if the user clicks on any item in table of contents. In addition, the user is able to store any information to their personal information area (as FOHM structures), simply by selecting a particular structure of information and the appropriate option to add it to their personal information area.
4. The system will manage the storing of the selected association to the personal linkbase.
5. The user is then able to add more data such as notes or comments to the data content in the personal information space, while the original remains unchanged.
6. The system will save all modification to the linkbases and data content which is represented in FOHM data object, in the personal information space.

3.3 Implementation

a-PIE is prototyped in Java as a web application under Apache Tomcat, using Auld Linky version 0.72 and Apache Axis. Apache Axis[1] is a java platform for creating and deploying web services applications. Java offers various advantages in comparison to other languages. Firstly, it can be used under different operating systems. Another important consideration is the availability of one of the most advanced frameworks to build Semantic Web applications, including a rule-based inference engine which is Jena[2]. a-PIE uses Jena, an opensource project, to manipulate RDF models and for a set of limited reasoning features.

The enriched documents are built by transformation from patterns catalogue through the presentation service. The metadata of FOHM associations and FOHM data objects are represented in RDF, while the ontologies for every service are represented in RDF/OWL.

4 Related Work

The Semantic Community Portal approach, provided by the Semantic Web Environmental Directory (SWED) [13], aims to overcome the limitations and problems with existing approaches to creating and maintaining web-based community information resources. The portal provides the system which enable third-party to reuse the information. The outstanding point for this directory management is the separation of data creation and storage from the publication. The Semantic Web technologies are used to enrich the basic SWED data records by various organisations. The data itself, written in RDF, is created and stored by the organisations on their own web sites. The directory organisation harvests the RDF files of organisations that are relevant to their particular area of interest. The directory organisation is able to add some additional specialist information themselves, or they might use their own vocabulary for categorising or describing the information. Therefore, this approach supports the reusability of information for the knowledge workers in organisations, not for an individual, and no adaptation supports.

The purpose of COHSE [14] is the integration of an open hypermedia architecture, especially the Distributed Link Service, with ontological services to support an architecture for the Semantic Web and provide the linking based on the concepts that appear in Web pages. Therefore, links are separated from the documents and links are manipulated separately from documents content. The documents are linked according to the metadata annotated to documents. The COHSE approach annotates the documents based on description logic and augments the documents with the annotations at browsing or

---

[1] http://ws.apache.org/axis/
[2] http://jena.sourceforge.net

reading time. However, there is no adaptation of the links and contents in this approach. In addition, COHSE does not provide the reusability of links for knowledge workers.

5 Summary

This paper describes a semantic adaptive information environment approach to support knowledge workers. The adaptive Personal Information Environment (a-PIE) based on the Semantic Web is proposed. The advantages of this approach are not only providing the adaptive information for particular users' needs, but also the reusability and sharebility of information for an individual. a-PIE provides an information environment for users in a community or organisation in which to browse or search information based on domain concepts defined by ontologies. The users are also able to manipulate their own information space by adding or deleting data or parts of information structures into their own information space. In addition, they can add personal information such as comments or notes to the existing data or information structures while the original data or information structures published remain unchanged. Moreover, it is possible to provide multiple aggregations and views of the same data in different contexts.

The adaptation, reusability and shareability of knowledge components in this system is achieved by using Semantic Web technology, and by storing separately the data, information structures, domain concept, context, behaviour, presentation and user information models. The data, information structure, context and behaviour are represented by FOHM object models and manipulated by using Auld Linky, a contextual link service. Ontologies are used to define common explicit relationships for domain concepts and to enrich data, information structures, presentation and user models. Moreover, the service-oriented framework is used to provide the loosely coupled and reusable software components.

Future work will focus on developing the ways for each community to create each model; data, structure, context, presentation and behaviour based on a service-oriented framework.

Acknowledgement

Our thanks to The Royal Thai Government for funding this work.

References

1. P. De Bra, L. Aroyo, and V. Chepegin, "The Next Big Thing: Adaptive Web-Based Systems," *Journal of Digital Information*, vol. 5, 2004.

2. D.T. Michaelides, D.E. Millard, M.J. Weal, and D.C. De Roure, "Auld Leaky: A Contextual Open Hypermedia Link Server," presented at Proceedings of the seventh Workshop on Open Hypermedia Systems ACM Hypertext Conference, Aarhus, Denmark, 2001.

3. D. Booth, H. Haas, F. McCabe, E. Newcomer, M. Champion, C. Ferris, and D. Orchard, "Web Services Architecture W3C Working Group Note," vol. 2005: W3C, 2004.

4. T. Berners-Lee, J. Hendler, and O. Lassila, "The Semantic Web," *Scientific American*, vol. 279, pp. 35–43, 2001.

5. T. R. Gruber, "A Translation Approach to Portable Ontology Specifications," *Knowledge Acquisition*, vol. 6, pp. 199–221, 1993.

6. "SKOS Core Guide," vol. 2005, A. Miles and D. Brickley, Eds.: World Wide Web Consortium, 2005.

7. P. Brusilovsky, "Adaptive Hypermedia," *User Modeling and User Adapted Interaction*, vol. 11, pp. 87–110, 2001.

8. C. Bailey, W. Hall, D. Millard, and M. Weal, "Towards Open Adaptive Hypermedia," presented at Proceedings of the second International Conference on Adaptive Hypermedia and Adaptive Web-Based System, Malaga, Spain, 2002.

9. O. De Troyer and S. Casteleyn, "Exploiting Link Types during the Conceptual Design of Web Sites," *International Journal of Web Engineering Technology*, vol. 1, pp. 17–40, 2003.

10. M. Bieber and J. Yoo, "Towards a Relationship Navigation Analysis," presented at Proceedings of the thirty-third Hawaii International Conference on System Sciences, 2000.

11. RDFConcept, "Resource Description Framework (RDF) Concepts and Abstract Syntax," G. Klyne and J. Carroll, Eds.: World Wide Web Consortium, 2004.

12. "OWL Web Ontology Language Overview," vol. 2005 June, D.L. McGuinness and F. v. Harmelen, Eds.: World Wide Web Consortium, 2004.

13. "Semantic Web Environmental Directory," http://www.swed.org.uk, 2004.

14. L. Carr, S. Kampa, D. De Roure, W. Hall, S. Bechhofer, C.G. Goble, and B. Horan, "Ontological Linking: Motivation and Analysis," *CIKM*, 2002.

A Multilayer Ontology Scheme for Integrated Searching in Distributed Hypermedia

C. Alexakos[1], B. Vassiliadis[3], K. Votis[1] and S. Likothanassis[1,2]

[1] Pattern Recognition Lab., Dept. of Computer Engineering and Informatics, University of Patras, Greece
{alexakos,botis}@ceid.upatras.gr
[2] Research Academic Computer Technology Institute, Greece
likothan@cti.gr
[3] Digital Systems and Media Computing Laboratory, Computer Science, Hellenic Open University, Greece
bb@eap.gr

Abstract. The wealth and diversity of information available in the internet or local hyper-media corpora has increased while our ability to search and retrieve relevant information is being reduced. Searching in distributed hypermedia has received increased attention by the scientific community in the past few years where integration solutions which rely on semantics are considered as efficient. In this paper we propose a new scheme for integrated searching in distributed hypermedia sources using a multi-layer ontology. Searching tasks are carried out in the metadata level, where information concerning hypermedia objects is published, managed and stored in the form of a scalable description of knowledge domains.

1 Introduction

The vision of a media-aware semantic web is one of the more exciting challenges faced by researchers of many scientific disciplines including those of hypermedia, information retrieval and distributed systems [7, 17]. In this context, searching in distributed and heterogeneous sources was always difficult but it seems that semantics may offer solutions to such unsolved problems [13].

The hypermedia community has already recognized the need for good search and query mechanisms in hypermedia systems [18, 20]. Halasz in [5] forecasted the need for both content-based and structure-based retrieval on hypermedia. This issue raises the need of a detailed description of hypermedia that is not only based on contents (represented by keywords) but also on semantic contents and contextual information. Early works have already recognized the need for managing distributed hypermedia but the appropriate technologies were missing [4]. In the following years, the need of knowledge

C. Alexakos et al.: *A Multilayer Ontology Scheme for Integrated Searching in Distributed Hypermedia*, Studies in Computational Intelligence (SCI) **14**, 75–83 (2006)
www.springerlink.com

representation combined with a set of rules and concepts has lead the evolution of ontologies as the main tool to describe data contents and their relations in modern information systems. An early definition presented in [15] describes ontologies as: "a hierarchically structured set of terms to describe a domain that can be used as a skeletal foundation for a knowledge base". An ontology's main feature is machine readability and understandability which in turn enable automatic cross-communication of different systems. This leads to increased platform independence as well.

Ontologically principled mechanisms and frameworks for hypermedia have been presented recently. Proposals such as the one of [11] suggest that semantics should be included in the conceptual modelling stage of hypermedia production. Topia [13] is an architecture for domain-independent processing of semantics and discourse into hypermedia presentations. In [6], a framework and a searching algorithm for locating distributed hypermedia in a P2P network is presented. Approaches such as semantically indexed hypermedia [16] and ontology-based linking [1] are also worth mentioning.

The introduction of the Semantic Web and its, nearly, unanimous approval has driven towards the representation of ontologies in semantics. For this purpose, a variety of semantic markup languages has been developed, based on the dominant XML standard. The most prominent ontology markup languages are DAML+OIL and its successor OWL, which is build on top of RDF Schema. The integration of hypermedia and semantic web technologies has been proposed both for standard [12] and open hypermedia configurations [3], for ontology matching [2] and management [9] while searching in distributed environments. Ontology schemes have been successfully used/designed for centralised management of images [10], XML document schemata [8], MPEG-7 semantically enriched content [19] and query processing in P2P [14]. Hot issues arise when integration is focused on the contents and not on the heterogeneous semantic metadata: lack of adequate knowledge representation methods, time delays in searching distributed hypermedia sources and restrictions due to the specificity of current ontology schemes used for searching.

One of the initial solutions to distributed hypermedia content integration was the use of a single global ontology scheme in order to describe all the contents of hypermedia systems. This centralised approach has proven to be inflexible since sources are exponentially increasing affecting query response times. On the other hand, content-based integration faces three distinct difficulties. First, the utilization of a single description scheme forces the usage of a specific ontology for the semantic annotation of hypermedia contents. Second, when the hypermedia information is provided by distributed servers, the amount of semantic descriptions is increasing according to content volume. Third, in some cases ontologies used are domain-specific. Specificity deters widespread use.

This work proposes the use of a multilayer ontology scheme in order to deal with the above mentioned issues in a distributed servers architecture that provides access to hypermedia content. The goal is to introduce a modular ontology based scheme that provides a scalable description of knowledge domains of hypermedia contents and structure. The advantage of the multilayer ontology lies in the fact that the representing scalable annotation of the hypermedia contents allows search engines to navigate easily and fast in the ontology-based index beginning from general described domains to more detailed description of the hypermedia contents.

The paper is structured as follows: Sect. 2 presents the three layer ontology scheme used for integration while Sect. 3 briefly discuses architectural issues of the proposed solution. Finally Sect. 4 concludes this paper.

2 The Multilayer Ontology Scheme

Our approach introduces a three layer semantic description of hypermedia contents: the Upper Search Ontology layer, that describes the basic concepts of the domains of knowledge of the content, a set of Domain Description Ontology layer that represent a more detailed description of each domain and the Semantic Metadata layer where the different semantic hypermedia description of the heterogeneous hypermedia servers lay. The proposed indexing scheme includes additional mapping information between the ontologies in the three layers providing the necessary information to search engines in order to navigate inside the ontology-based index. Also, the model utilizes a lexical ontology that comprises a set of lexical and notational synonyms reinforcing the searching among the various ontology instances of the hypermedia content representation. The three layer ontology scheme is depicted in Fig. 1.

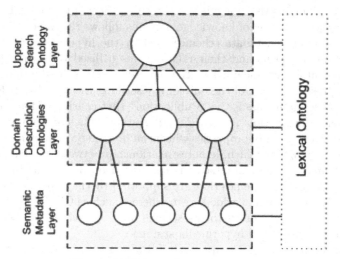

Fig. 1. Multilayer Ontology Scheme

The upper layer of the ontology scheme is comprised of a single ontology describing the main knowledge domains that represent the contents of the integrated hypermedia systems. This layer is primarily used as a "catalogue" for the type of contents and the domain of knowledge that are integrated. Domains are specified in classes and subclasses providing a hierarchical model presenting all the knowledge fields that are included in the hypermedia contents of the distributed servers. There is also a number of properties denoting the relationship between classes.

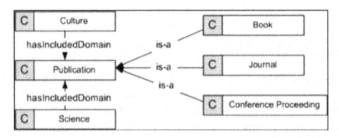

Fig. 2. Example of an Upper Search Domain

Figure 2 presents an example of an upper search ontology consisting of two main classes "Culture", "Science" and "Publications". The "Publications" class contains three subclasses the "Books", "Journals" and "Conference Proceedings". There is also a connection of the type "hasIncludedDomains" denoting that the classes "Culture" and "Science" include publications.

In the middle layer, lays a set of ontologies for the detailed descriptions of each of the domains that included in the Upper Search Ontology Layer. The semantic representation of knowledge domains follows the same level of detail with the semantic metadata schemes used by the hypermedia servers. This ensures that all terms and their relationships utilized by each hypermedia server separately are included in the ontology scheme in the middle layer. Figure 3 presents a part of the ontology describing the knowledge domain of Journals. The main class is the "Publication" that consists of four subclasses "Publisher", "Institution", "Article" and "Author". Each of these subclasses is characterized by a set of properties which can be either a simple data property or an object property which de-notes relationship between two classes.

The Semantic Metadata Layer is comprised of all the variant semantic annotated metadata schemes that are supported by the hypermedia systems. These schemes may follow wide known schemes such as CIDOC-CRM, Dublic Core or any other custom ontology scheme used by each of the search mechanisms in the integrated hypermedia systems.

As presented before, our approach introduces three different layers of semantic descriptions. In this context, a complete description of different hypermedia contents is possible. What is still missing is the association of the semantic information represented in these three layers. This association promotes an

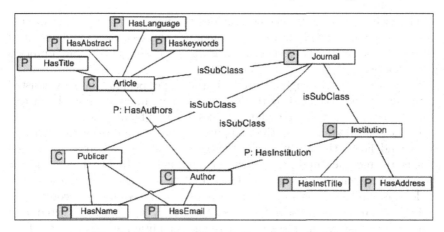

Fig. 3. Example of Domain Description Ontology

integrated description of enterprise-relevant knowledge that can be machine-readable. A methodology is required to address the map-ping needs between the different implemented ontology layers providing the aforementioned semantic information.

The first step in this methodology concerns the mapping of the domain ontologies of the middle layer to the Upper Search Ontology. This mapping is possible by affiliating the main classes of domain ontologies to the corresponding classes of the Upper Search Ontology. In the previous examples, the mapping can be utilized by associating the class "Journals" of the Upper Domain Ontology to the class "Journal" of the domain search ontology.

The second step in our methodology regards the need for mapping between different domain description ontologies. This mapping is necessary, since each domain ontology represents semantics of different knowledge domains. Given that layer functionalities differ significantly, similar differences may be present to the corresponding ontologies. Since integration needs require that information is seamlessly passed among the different layers, domain ontology mapping is absolutely necessary.

The final step in our methodology introduces the mapping of the domain search ontologies to the semantic information related with the actual description of hypertext content in the different hypermedia systems. This mapping is related to the description, in ontology terms, of the hypermedia content (e.g. which server provides what). This semantic information stems from the semantic metadata description of the hypermedia content and has to be mapped to the corresponding classes and properties of the relevant domain description ontology.

This methodology leads to the desirable unification of the Upper Search Ontology with the domain description ontologies and the semantically described hypertext contents, and thus contributes to the creation of a flexible and reusable ontology metadata scheme.

The proposed scalable scheme provides the necessary supporting mechanism to a search engine to navigate through the ontology terms and instances faster and efficiently. The basic concept is the separation of the search process in three steps.

In the first step, the engine searches in the Upper Search Ontology assisted by the synonyms extracted by the lexical ontology aiming to find the proper general domain (or domains) where the results may be included.

In the second step, using the mapping information between the first and the second layer, the engine seeks the appropriate domain description ontologies to find the appropriate structural ontology terms that describe the corresponding semantic metadata structure. The linguistic information from the lexical ontology is also reinforcing the search process in the second step.

In the final step of the search process, the search engine has extracted from step two the basic structural elements of the semantic metadata that will be used in the hypermedia content searching. Utilizing the mapping information between the second and the third layer, the engine constructs the appropriate query for each of the distributed hypermedia servers. Each query is consisted of terms that are included in the semantic metadata in each hypermedia server.

3 Architectural Issues of the Proposed Solution

The proposed multilayer ontology scheme can decrease the time of search by reducing the number of the distributed hypermedia servers that accept the search query. Through the first two steps of the search process, the search will result in a number of domain description ontologies that correspond to a specific number of servers; this concludes the execution of the query in these servers, not to all servers in the distributed hypermedia net-work.

This modular approach also contributes to ontology simplicity and integration ease of hypermedia servers with different semantic metadata. The multilayer scheme allows the import of new type of semantic metadata by initializing the mapping information with the corresponding Domain Description Ontology Layer. Also, in a similar way the accession of the Do-main Description Ontology in the second layer can be done with a mini-mum cost of manual work. The suppleness of the multilayer ontology model provides a simple mechanism to index hypermedia content in the distributed servers.

The general representation of knowledge that is provided by the layers of the ontology scheme empowers the search engine to search in each do-main of knowledge which the hypermedia content belongs to. The capability of adding new domains and new servers storing hypermedia content does not restrict the search process in certain domains.

The enforcement of the above presented multilayered ontology scheme requires a specific architecture that makes it possible to integrate the different semantically annotated hypermedia data residing in different servers in a flexible and interoperable way. Our approach introduces a framework for

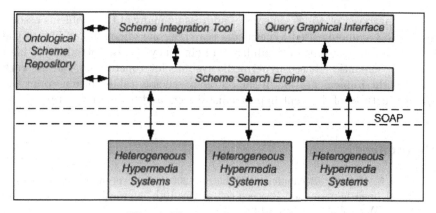

Fig. 4. The associate architecture

developing the multilayer ontology scheme and managing the search queries
from the users. This framework, presented in Fig. 4, involves the following
functional elements:

- The *Ontology Scheme Repository* where the ontologies contained in the
  multilayer scheme are described using XML-based ontology representation
  standards.
- The *Scheme Integration Tool* that provides the mechanism and the graphical interface for composing the mapping information between the ontologies
  of the layers.
- The *Query Graphical Interface* providing the appropriate search forms to
  the user of the integrated system.
- The *Scheme Search Engine* is the platform which accepts the queries
  from the Query Graphical Interface and initiates the ontology-based search
  process.

4 Conclusions

Searching in distributed hypermedia resources is challenging problem and it
seems that semantic web technologies, such as ontology-principled repositories, offer a viable solution.

In this work we presented a three-layer ontology scheme that enables efficient search of hypermedia objects stored in a distributed server environment.
The main advantage of the proposed architecture is that it provides a flexible
way for the semantic description of knowledge stored in different hypermedia
systems and is already annotated with different semantic metadata standards.
The multilayer ontology scheme and the associated architecture provide an integrated approach towards achieving integration of different and distributed
hypermedia systems.

Our proposal provides a means of addressing many of the problems associated with commonly used hypermedia searching paradigms by adding to the overall system in terms of flexibility and efficiency. It also introduces a high degree of semantic – based hypermedia systems interoperability. Future work includes the implementation of the ontology scheme, taking into account the specific features of different hypermedia objects and testing using real data.

References

1. Carr, L., Hall, W., Bechhofer, S., Goble, C. (2001). Conceptual linking: ontology-based open hypermedia. Proceedings of the 10th international conference on World Wide Web, pp. 334–342.
2. Doan, A., Madhavan, J., Dhamankar, R., Domingos, P., Halevy, A. (2003). Learning to match ontologies on the Semantic Web. The VLDB Journal – The International Journal on Very Large Data Bases, Vol. 12(4), pp. 303–319.
3. Dolog, P., Henze, N., Nejdl, W. (2003). Logic-Based Open Hypermedia for the Semantic Web. In Proc. of International Workshop on Hypermedia and the Semantic Web, Hypertext 2003 Conference, Nottingham, UK.
4. Goose, S., Dale, J., Hill, G., de Roure, D., Hall, W. (1996). An Open Framework for Integrating Widely Distributed Hypermedia Resources. Proceedings of the 1996 International Conference on Multimedia Computing and Systems (ICMCS '96), pp. 0364.
5. Halasz, F. (1988). Reflections on NoteCards: Seven Issues for the Next Generation of Hypermedia Systems. Communications of the ACM, Vol. 31, No. 7, July, pp. 836–852
6. Larsen, R.D., Bouvin, N.O. (2004). HyperPeer: searching for resemblance in a P2P network. Proceedings of the 15th ACM conference on Hypertext & Hypermedia, pp. 268–269
7. Li, WS., Candan, K.S. (1999). Integrating content search with structure analysis for hypermedia retrieval and management. ACM Computing Surveys (CSUR), Volume 31, Issue 4es, Article No. 13.
8. Lu, E.J.L., Jung, Y.M. (2003). XDSearch: an efficient search engine for XML document schemata. Expert Systems with Applications, 24, pp. 213–224.
9. Maedche, A., Motik, B., Stojanovic, L. (2003). Managing multiple and distributed ontologies on the Semantic Web. The VLDB Journal – The International Journal on Very Large Data Bases, Vol. 12(4), pp. 286–302.
10. Mezaris, V., Kompatsiaris, I., Strintzis, M.G. (2004). Region-based Image Retrieval using an Object, Ontology and Relevance Feedback. Eurasip Journal on Applied Signal Processing, Vol. 2004 (6), pp. 886–901.
11. Montero, S., Diaz, P., Aedo, I., Dodero, J.M. (2003). Toward Hypermedia Design Methods for the Semantic Web. Proceedings of the 14th International Workshop on Database and Expert Systems Applications, pp. 762.
12. Nanard, M., Nanard, J., King, P. (2003). IUHM: a hypermedia-based model for integrating open services, data and metadata, Proceedings of the 14th ACM conference on Hypertext and hypermedia, pp. 128–137.
13. Rutledge, L., Alberink, M., Brussee, R., Pokraev, S., van Dieten, W., Veenstra, M. (2003). Hypermedia semantics: Finding the story: broader applicability of

semantics and discourse for hypermedia generation. Proceedings of the 14th ACM conference on Hypertext and hypermedia, pp. 67–76.

14. Stuckenschmidt, H., Giunchiglia, F., van Harmelen, F. (2005). Query Processing in Ontology-Based Peer-to-Peer Systems. Ontologies for Agents: Theory and Experiences, Whitestein Series in Software Agent Technologies.

15. Swartout, B., Patil R., Knight K., Russ T. (1996). Toward distributed use of large-scale ontologies. In Proceedings of the 10th Knowledge Acquisition for Knowledge – Based Systems Workshop.

16. Tudhope, D., Cunliffe, D. (1999). Semantically indexed hypermedia: linking information disciplines. ACM Computing Surveys (CSUR), Volume 31, Issue 4es, article No. 4.

17. Van Ossenbruggen, J., Nack, F., Hardman, L. (2004). That Obscure Object of Desire: Multimedia Metadata on the Web, Part 1. IEEE MultiMedia, Vol. 11(4), pp. 38–48.

18. Vitali, F., Bieber, M. (1999). Hypermedia on the Web: what will it take?. ACM Computing Surveys (CSUR), Volume 31, Issue 4es, article No. 31.

19. Westermann, U., Klas, W. (2003). An Analysis of XML Database Solutions for the Management of MPEG-7 Media Descriptions. ACM Computing Surveys (CSUR), Vol. 35(4), pp. 331–373.

20. Wiil, U.K., Nürnberg, P.J., Leggett, J.J. (1999). Hypermedia research directions: an infrastructure perspective. ACM Computing Surveys (CSUR), Volume 31, Issue 4es, Article No. 2.

htmlButler – Wrapper Usability Enhancement through Ontology Sharing and Large Scale Cooperation

Christian Schindler, Pranjal Arya, Andreas Rath, and Wolfgang Slany

Institute of Software Technology, Graz University of Technology
{cschindl,parya,arath,wsi}@ist.tugraz.at

Abstract. The htmlButler project aims at enhancing the usability of visual wrapper technology while preserving versatility. htmlButler will allow, for an untrained user who has only the most basic web knowledge, to visually specify simple but useful wrappers and, for a more tech-savvy user, to visually or otherwise specify more complex wrappers. htmlButler was started 2005/2 and is based on visual wrapping technology research carried out in the Lixto project since 2000. What is new in htmlButler is that (a) the application is entirely server based, the user accessing it through his or her standard browser, (b) because of the centralized wrapper configuration and processing, the knowledge about popular wrappers can be leveraged to facilitate the specification of wrappers for new users, and (c) users can contribute narrow and precise ontologies that help the system in recognizing potential meaning in web pages, thereby alleviating the complexity of future wrapper configurations

1 Introduction

Use cases in different vertical market domains suggest that end users (e.g., quality managers in the automotive domain, but also private end users) are eager to wrap and aggregate Web data, e.g., to notify themselves of changes in particular parts of pages of interest. However, specifying what to wrap with current interactive wrapping technology is a difficult task for untrained users. Note that the same users have little trouble communicating what they want to be wrapped to other humans, and wrappers can usually wrap that information once it has been correctly specified. The problem therefore lies in the user interface. We would like to investigate whether the communication between human users and wrapper technology can be improved through the use of high-level semantic concepts. Our aim is to make wrapper technology understand high-level concepts that human users might want to employ when wrapping web pages. Ontologies that provide systematic, computer oriented representations of real world semantic concepts and relations between them can be used to store and reason about such semantic concepts. An opportunity

C. Schindler et al.: *htmlButler – Wrapper Usability Enhancement through Ontology Sharing and Large Scale Cooperation*, Studies in Computational Intelligence (SCI) **14**, 85–94 (2006)
www.springerlink.com

that arises in the realm of the Internet and in particular when we aim at widespread adoption of wrapper technology is that user communities can cooperate in building up shared ontologies. A wider audience will benefit from such an effort, thus building up a positive feedback loop with more and more users contributing to and mutually benefiting from an increasingly intelligent ontology based Web wrapper.

2 Previous Technology

htmlButler is based on the Lixto set of tools that allow application developers to implement such processes without the need for manual coding. The Lixto Visual Wrapper generation tool is based on a new method of robustly identifying and extracting relevant content parts of HTML documents and translating the content to XML format [1]. Lixto wrappers are embedded into an information processing framework, the Lixto Transformation Server [2]. The Lixto Transformation Server enables application developers to format, transform, integrate, and deliver XML data to various devices and applications. Using ontologies in wrapper specification as well as the creation of shared ontologies has been studied previously [3, 4, 5]. However, we are not aware of attempts to combine these two approaches. Also, while many ontology projects eventually succeed in the task of defining upper domain ontology, populating the third level, what is called the specific domain ontology, is the actual barrier that very few projects could overcome so far. We will investigate how cooperatively created, shared ontologies with corresponding new user interface paradigms could be used to overcome this barrier.

3 htmlButler

To satisfy a low inhibition threshold, we postulated the following criteria:

No Installation: The user need not install any new application or plug-in for the browser. This eliminates the concern of downloading a potentially harmful program.

Easy Configuration: It is not required to have any programming or web technology knowledge such as HTML or HTTP to create a wrapper. Simple wrappers can be configured entirely in a visual interactive way. Should similar wrappers or wrapped pages exist, then the system will suggest wrapper configuration accordingly.

Usability: The user just has to enter an URL of a website, his email address and the frequency or schedule the extracted information should be sent to him. Alternatively to an exact URL, the system also will accept keywords that return a list of sites found by some search engine like google. Navigation and login protocols needed for certain non persistent URLs or password protected pages will also be available.

Resistance: The generated wrappers will be resistant to slight website changes so that users do not have to reconfigure wrappers frequently.

Easy Maintenance: Reconfiguration and enhancing of existing wrappers is done interactively with suggestions made by the system, requiring user approval.

Appropriate User Levels: A user will be able to make the most out of his skills.

3.1 No Installation

htmlButler is a Web based application and hence the user does not have to download and install any files. As a result of this htmlButler can be used on computers with restricted access rights like on public Internet terminals or in Internet cafes. Furthermore for using htmlButler only the authorization of executing JavaScript code inside a standard browser (eg. Microsoft Internet Explorer) is needed. All wrapping and processing takes place on the server side and hence the speed of the user's machine is not slowed down and wrappers are executed independently of the user's computer.

3.2 Easy Configuration and Usability

Since htmlButler can also be used by people with limited technological knowledge, easy configuration is a crucial aspect. When a user has chosen a website from which specific data should be wrapped, htmlButler searches its database for existing wrappers that have been created for that page and suggests which data has been extracted by other users of the system. It is also possible for a user to select another area to extract. The user than enters the notification details and the data that he extracted will be sent to him according to user specified criteria, e.g. whenever changes occur, or according to some notification schedule.

3.3 Resistance and Maintenance

The organization of websites changes frequently, so the wrappers have to change as well but wrappers that have been created by the htmlButler application are resistant to slight website changes (e.g. html tags are added, small layout changes)[1].

3.4 Appropriate User Levels

At the novice level, he will just be using the notification service for change. At the second stage, he would be able to enhance the simpler wrappers of the novice user. At the third and the final level, he would have the liberty to create the ontology and semantic concepts for a particular domain. Basically there are three different categories of htmlButler users. First there is the novice user who is just familiar with web browsing and has no knowledge about wrapper and internet technology. Second the advanced user who has some knowledge about web technology and is able to extend the wrapper scenarios of the novice user which have been created almost automatically by the htmlButler software. Finally there is the tech-savvy user who has knowledge about ontologies and se-mantics and can submit them to the htmlButler system to tune the creation process and make the htmlButler system smarter and the wrapper creation easier.

3.5 htmlButler in Action

One sample usage scenario is the generation of XMLTV formatted TV program listing information. XMLTV can then be used as a data-source for TV program guides. One simple way to use this is as a website of a TV Guide. Furthermore, it is possible to use XMLTV for scheduling recordings or displaying detailed infor-mation for the current show. However, by now there is no existing solution that works perfectly. The current state is that such programs are quite slow (due to the slow grabbing process), do not instantly update TV listings if the TV program changes and many applications are not yet very user-friendly. With our application we plan to acquire the data and maintain it continuously. We would also be in the position to send notification about changes in the TV program thus allowing automatic rescheduling of recording information because no human intervention in the data maintenance part is needed. Fetching data from a new TV station would be easy with the wrapper. Also there would be instant response to changes on the data supplier's website by the community. We can have a number of user categories and community cooperation. For example, a tech-savvy user can define rules for creating the initial XML database. This user is also responsible for defin-ing what information belongs to each element and also defines rules to compute the different attributes. The second user category is the advanced user. The main role of this user category is to provide data about new TV station sites by provid-ing the locations from where to wrap the data. The server will then automatically generate an XML file according to the rules and make it publicly available. The file can then be retrieved via HTTP. Other interfaces (like SQL) are also possible.

4 Semantic Supported Wrapper Generation

Narrow but detailed ontologies can enhance the process of creating wrappers in supporting the user with intelligent suggestions about extractable content of the web page the user is interested in. For instance in the system an ontology describing TV shows and typical information found therein including the names of popular artists and directors or series help a user who wants to add a new TV station in configuring the wrapper needed to map the listings to the XMLTV format, much like a human operator would recognize the various entries found on such a page. htmlButler thus emulates the human understanding of what a website is all about based on sample content. A more advanced user could semi-automatically add new or missing concepts to the ontology based on his or her current needs to make the wrapper generation easier, from which other users might profit in the future. Already working wrappers can be reused and enhanced and work as a base for more complex usage scenarios. The ontology can be used for verifying the results of the created wrapper and adjust the wrapper to changes in the web site if for instance a column is inserted to a TV listing table and the program description has thereby changed place.

5 Shared Ontologies

So called extraction ontologies can help to make the wrapper configuration process easier [6]. Although the creation of ontologies is usually hard work, an expert can quickly create reasonably good extraction ontologies for a narrow domain of interest. In our project we aim at letting non-expert users combine and manipulate public contributions to centrally maintained shared ontologies, thereby lifting some of the weight from individual contributors by being able to make use of existing extraction concepts and aggregation relations. Even more, incompletely contributing narrow domain ontologies will still make sense since a specific goal that is of interest to the contributing user will be served. For instance, assume that the user is interested in wrapping an ordered list of names of book authors from a page. The user would directly select the column using the mouse, and the wrapper would suggest either to notify the user of changes in that column (the first choice if that was what most users chose), or to give the column the possible name "list of authors" for further aggregation or selection, by comparing the strings found in the column with a list of concept instances found in the wrapper's ontology that contains names of known and popular authors. The latter could have been contributed by some other, volunteer user that extracted that information for his own wrapping purposes from some public source or some source available for private use, e.g. the Amazon web site or its web service, or some public library catalog, or Wikipedia entries about authors. The wrapper could also suggest concept types such as "list of family names", "list of names", "list of strings", "this particular list of entries", or simply "generic list", in decreasing order of specificity, respectively by likelihood and usefulness through statistics on number of recent choices of these concepts by other users. Users would also have the choice to give a new name to the list, but in this case the wrapper would suggest already existing similar concepts based on a thesaurus or foreign language dictionary lookup and other similarity heuristics, and users would be given the choice of enlarging or editing existing concepts instead of creating new ones. For the latter a system of versioning of concepts would be needed in order not to break existing wrappers of users having used a previous version of the concept, including a notification of these changes to the previous user. Also, sub-concepts (e.g. "list of classic German authors") or new super-concepts or foreign-language varieties could be easily created in analogous manners.

In a similar fashion, relations between concepts as described in the shared ontologies could be used to suggest likely aggregation schemes (for the composition of queries spanning several Web pages), again using a variety of heuristics, e.g. statistical information regarding what aggregates where chosen by other users who used the chosen concepts, and what concepts the present user chose on this or other pages that semantically bear relations to the content selected by him on the present page. A related problem that immediately arises with shared ontologies is quality. However, we will experimentally investigate whether sacrificing quality for usefulness will be sufficient for efficient extraction purposes. To somehow recover practical quality, we will investigate self-correction systems where users, e.g. can rate the usefulness of ontologies contributed by others. These ratings can then be used to influence the way the wrapper will suggest tokens that should be wrapped on similar occasions. We will therefore study existing tools for their suitability for non-expert, casual use ontology contribution, editing, and assessment. One thing we will need to address is supporting the group of actively contributing users and of passively consuming users (that will likely overlap according to situation) in the discussion and decision-making process required by ontology building. We will study whether this can be achieved by groupware tools that can be freely self-organized such as WikiWikiWebs [7] that are considered efficient knowledge management and knowledge sharing tools. This would greatly facilitate free-format organizational communication, allowing volunteer participation in organizing FAQ lists, documentation, tutorials, how-to guidelines, feature brainstorming boards, ontology annotation repositories, and discussion places for contributed ontologies, all in a persistent manner that encourages neutral-point-of-view consensus finding as, e.g. in the Wikipedia project [8].

As the aim of making wrappers understand users based on intentions and meaning falls nothing short of many unsolved artificial intelligence goals and thus may seem very difficult to achieve even from a long-term perspective, we would like to point out that realizing already a small part only of the proposed research program will yield non-trivial insights in what would be needed achieve a better communication between computers and humans.

6 htmlButler Status Quo

Technically, htmlButler is almost completely server based – only selecting the area of interest on the web page, which should be wrapped, is done on the client side using JavaScript. At the htmlButler start page, the user will be

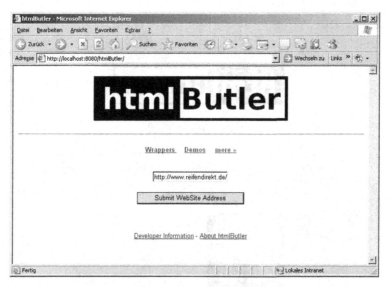

Fig. 1. Submission of the URL which should be wrapped

prompted for login or create a new user with the information, where to send
the wrapping output and the default frequency of wrapper invocation. The
user enters the webpage's address (see Fig. 1). This URL is submitted to the
server, the content of this URL is fetched and processed in a certain way and
sent back to the user's browser embedded into a frameset (see Fig. 2). The user
can then select the area of interest and submit this selection to the server. At
the server side the selection is processed and now, by means of implicitly stored
rules and semantics about the structure of web pages, a wrapper is generated
and scheduled for invocation. At the actual implementation the user is not
able to alter or enhance these rules. As the htmlButler project reaches its next
stage, it would be possible for people to log in as a super user (or an enhanced
user with more liberties to create complex wrappers). It will be a challenge
to improve the user interface in that way to enable the user enhancing or
submitting new concepts for the wrapper generation. htmlButler combines
two basic mechanisms for data extraction – tree and string extraction. For tree
extraction the elements are identified with their corresponding tree path and
possibly with some properties of the elements themselves. For string extraction
the leaves of the HTML parse tree are examined in detail for changes.

7 Conclusion

We are interested in studying the preconditions necessary to allow the use
of knowledge representation formats in wrapper specification user interfaces.
We also study means to allow users to easily share their knowledge through

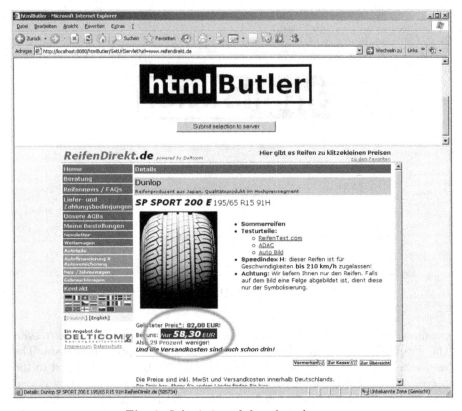

Fig. 2. Submission of the selected area

existing knowledge representation formats used in other semantic Web applications in order to leverage their contribution efforts so that economy of scale effects can take place. Standards that allow the interoperability of different ontologies will be important to allow the integration of existing ontologies. We will thus investigate how such standards can be brought to use in wrapper technology, both for concept recognition as well as for relations between concepts for the aggregation of wrapped data. A preliminary problem that is however fundamental to the success of our research agenda consists in the necessity to make wrapper user interfaces accessible to a wider audience, triggering economy-of-scale effects in the long run. We will therefore research usability aspects for a widespread acceptance of wrapper technology.

When visually specifying a wrapper, the system will suggest most likely concepts for further aggregating the data based on knowledge about syntactic and semantic concepts in the context of the page. The set of concepts covers basic concepts for wrapping (e.g. syntactic concepts such as time or date formats) and concepts at a higher aggregation level (e.g. names of popular actors or types of TV programs in various languages) that were defined by other users on similar pages. Thus the system is imitating to a certain extend the behaviour of an experienced human user. We believe it is necessary to investigate the particular aspects of ontology sharing and ontology usage in wrapper applications. We will in particular investigate whether the interaction between untrained users and wrapper generators can be enhanced by community-shared, -maintained, and -validated ontologies.

The outcome of this research should be a set of new methods, algorithms, and heuristics that allow the system to efficiently use shared ontologies in an interactive wrapper specification process, resulting in enhanced usability. Moreover, methods for validation and manipulation of shared ontologies as well as interfacing between different existing ontologies will be developed. Additional research will concern guidelines for both the ontology-based wrapping process as well as the management of shared ontologies.

The htmlButler project mainly deals with usability aspects in the context of wrapper construction by untrained users. Defining a proper user interface with only limited complexity but sufficient expressiveness is the key challenge of this task. We are convinced that the use of shared ontologies can be a vehicle to achieve this goal.

Acknowledgements

The htmlButler project is funded by the Austrian FFG [9] under the FIT-IT [10] program line as a part of the NextWrap project.

References

1. R. Baumgartner, S. Flesca, G.G.: Visual web information extraction with lixto. In: Proceedings of VLDB. (2001)
2. M. Herzog, G.G.: Infopipes: A flexible framework for m-commerce applications. In: Proceedings of TES workshop at VLDB, (2001)
3. M. Hatala, G.: Global vs. community metadata standards: Empowering users for knowledge exchange. In: Proceedings of the First International Semantic Web Conference. (2002)
4. M. Missikoff, X.W.: Consys – a group decision-making support system for collaborative ontology building. In: Proceedings of Group Decision & Ne-gotiation 2001 Conference. (2001)
5. M. Stonebraker, J.H.: Content integration for e-business. In: ACM Sigmod Conference. (2001)
6. D.W. Embley, C.T., Liddle, S.: Automatically extracting ontologi-cally specified data from html tables with unknown structure. In: Proceedings of the 21st International Conference on Conceptual Modelling. (2002)
7. WikiWikiWeb: http://en.wikipedia.org/wiki/WikiWikiWeb, visited May 2005 (2005)
8. Wikipedia: The Free Encyclopedia – http://wikipedia.org, visited May 2005 (2005)

9. FFG: Österreichische Forschungsförderungsgesellschaft mbH – http://www.ffg.at, visited May 2005 (2005)
10. FIT-IT: Forschung, Innovation, Technologie – Informationstechnologie – http://www.fit-it.at, visited May 2005 (2005)

A Methodology for Conducting Knowledge Discovery on the Semantic Web

Dimitrios A. Koutsomitropoulos, Markos F. Fragakis, Theodoros S. Papatheodorou

University of Patras, School of Engineering, Computer Engineering and Informatics Dpt. High Performance Information Systems Laboratory Building B, 26500 Patras – Rio, Greece
kotsomit@hpclab.ceid.upatras.gr,
fragakis@ceid.upatras.gr,
tsp@hpclab.ceid.upatras.gr

Abstract. One of the most prominent features that the Semantic Web promises to enable is the discovery of new and implied knowledge from existing information that is scattered over the Internet. However, adding reasoning capabilities to the existing Web infrastructure is by no means a trivial task. Current methods and / or implementations do not seem to be declarative and expressive enough to provide the kind of reasoning support that the Semantic Web users will benefit from. In this paper we propose a methodology based on which, the user can construct and pose intelligent queries to Semantic Web documents in an intuitive manner, without prior knowledge of the document's contents or structure. This methodology is implemented through a knowledge discovery prototype interface that relies on an existing inference engine found suitable for this task.

1 Introduction

One of the most important features promised by the Semantic Web is the capability of reasoning and knowledge discovery. Several solutions have been proposed focused on founding or improving reasoning tasks since the advent of the Semantic Web. Ontology description languages, inference systems and engines, as well as implementations with reasoning capabilities have been developed in order to improve information retrieval and enable knowledge discovery, if possible. Examining existing approaches reveals that no well established and standardized methodology is followed in general. In addition, the expressiveness achieved varies, not always fulfilling the expressiveness needs of the Semantic Web. Finally, current approaches for composing and performing intelligent queries do not seem to be satisfyingly declarative. Most of the time, the user is burdened with the task of collecting the appropriate information needed to construct the query.

D.A. Koutsomitropoulos et al.: *A Methodology for Conducting Knowledge Discovery on the Semantic Web*, Studies in Computational Intelligence (SCI) **14**, 95–105 (2006)
www.springerlink.com   © Springer-Verlag Berlin Heidelberg 2006

In this paper we propose a methodology for conducting knowledge discovery on the Semantic Web. This methodology consists of three phases: First, the selection of the appropriate logical formalism that will enable the use of existing AI tools and reasoners to perform inferences on Semantic Web documents. Second, the identification of some key aspects that need to be taken into account in order for our method to be suitable for a Web environment. Third, the selection of a specific inference engine that meets the criteria posed in the previous phase.

An integral part of our methodology is the Knowledge Discovery Interface (KDI). The KDI is a web application that has been developed as an implementation of the decisions and criteria developed during the three phases. As such, its target is threefold:

- To be *expressive* enough, in order to allow for powerful inferences on ontology documents.
- To be *declarative*, being independent of the specific ontology schema or contents.
- To be *intuitive*, by aiding the user to compose his query in a user-friendly manner and allowing transparent query composition.

The rest of this paper is organized as follows: In Sect. 2 we review and discuss some previous work on information retrieval and knowledge discovery. Then, in Sect. 3, we propose our methodology for knowledge discovery on the Semantic Web. The KDI is presented in Sect. 4, followed by some experimental results that demonstrate its capabilities. Finally, Sect. 5 summarizes our conclusions.

2 Approaches for Reasoning on the Web

Even though the idea of the Semantic Web has only recently begun to standardize, the need for inference extraction and intelligent behaviour on the Internet has long been a research goal. As expected, there have been some efforts in that direction. Such efforts include ontology description languages, inference engines and systems and implementations, based on them.

Knowing the constraints of knowledge discovery in a random environment like the Internet, and taking in to account the advantages of information retrieval, recent research has tried to combine these two approaches. OWLIR [18] for instance, is a system conducting retrieval of documents that are enriched with markup in RDF, DAML+OIL or OWL. A text editing and extraction system is used to enrich the documents, based on an upper level ontology. This extra information is processed by a rule-based inference system. Search is conducted using classical retrieval methods; however, the results are refined using the inference system results.

The TAP framework [9] seeks as well to improve the quality of search results by utilizing the semantic relationships of web documents and entities.

However, no inference takes place here. Instead, the RDF/OWL documents are treated as structured metadata sets. These sets can be represented as directed graphs, whose edges correspond to relations, and vertices correspond to existing internet resources.

The growth and maintenance of a knowledge base is a strenuous procedure, often demanding a great extent of manual intervention. The Artequakt system [1] tries to overcome this obstacle following an automated knowledge extraction approach. Artequakt applies natural language processing on Web documents in order to extract information and uses CIDOC-CRM as its knowledge base conceptual schema. Nevertheless, it should be noted that no inference – and thus knowledge discovery – takes place.

The Wine Agent system [15] was developed as a demonstration of the knowledge discovery capabilities of the Semantic Web. This system uses a certain domain ontology written in DAML+OIL/OWL and performs inferences on it. The Wine Agent employs a first order logic theorem prover (JTP).

The need for formal querying methods with induction capabilities, has led to DQL [5], as well as OWL-QL [6]. DQL and OWL-QL play an important role in terms of interoperability, expansion and enablement of intelligent systems on the Semantic Web. Nevertheless, they do not provide a direct answer to the knowledge discovery issue. Instead, they serve mainly as communication protocols between agents.

3 An Inference Methodology for the Semantic Web

In this section we propose the basic characteristics that a concrete and declarative method for designing intelligent queries should have in order to be appropriate for the Semantic Web environment. Having decided on an appropriate logical formalism, the resulting methodology relies on an inference engine that should meet certain criteria and be suitable for this purpose. By combing low-level functions of such an engine, our methodology aims at helping the user construct his query in a declarative manner and enables expressive intelligent queries to web ontology documents.

3.1 Choosing the Appropriate Formalism

Choosing an underlying logical formalism for performing reasoning is crucial, as it will greatly determine the expressiveness to be achieved. In this subsection we will attempt to examine some available formalisms, as well as a number of existing tools for each of them.

Description Logics (DL) form a well defined subset of First Order Logic (FOL). It has been shown [12] that OWL DL can be reduced in polynomial time into $SHOIN(D)$, while there exists an incomplete translation of $SHOIN(D)$ to $SHIN(D)$. This translation can be used to develop a partial, though powerful reasoning system for OWL DL. A similar procedure is followed for the reduction of OWL Lite to *SHIF*(D), which is completed in polynomial time as well. In that manner, inference engines like FaCT and RACER can be used to provide reasoning services for OWL Lite/DL.

A fairly used alternative are inference systems that obtain reasoning using applications based in **FOL (theorem provers)**. Such systems are Hoolet, using the Vampire theorem prover, Surnia, using the OTTER theorem prover and JTP [7] , used by the Wine Agent. Inference takes place using axioms reflecting the semantics of statements in OWL ontologies. Unfortunately, these axioms often need to be inserted manually. This procedure is particularly difficult not only because the modeling axioms are hard to conceive, but also because of their need for thorough verification. In fact, there are cases where axiom construction depends on the specific contents of the ontology [15].

Another alternative is given by **rule based reasoning systems**. Such systems include DAMLJessKB [16] and OWLLisaKB. The first one uses Jess rule system to conduct inference on DAML ontologies, whereas the second one uses the Lisa rule system to conduct inference on OWL ontologies. As in the case of theorem provers, rule based systems demand manual composition of rules that reflect the semantics of statements in OWL ontologies. This can also be a possible reason why such systems can presently support inference only up to OWL Lite.

DLs seem to constitute the most appropriate available formalism for ontologies expressed in DAML+OIL or OWL. This fact also derives from the design process of these languages. In fact, the largest decidable subset of OWL, OWL DL, was explicitly intended to show well studied computational characteristics and feature inference capabilities similar to those of DLs. Furthermore, existing DL inference engines seem to be powerful enough to carry out the inferences we need.

3.2 Suitability for the Web

Research regarding the use of ontologies and Description Logics for semantic matchmaking of Web Services descriptions [8, 17, 20] as well as other sources [19] has led to the recognition of a common query method, using subsumption relationships between concepts. According to this method, the query is modeled as a new concept, using the Description Logic constructors, and then classified in the hierarchy. Subsumption relationships then determine the absolute and relevant answers to the query. Furthermore, every instance can be modeled as an atomic concept, so queries demanding use of instances can be conducted the same way [14]. The method just described, could be characterized as *taxonomic* since it is entirely based on the functions provided by the TBox of a knowledge base.

Although artificial class creation seems adequate for the matchmaking of Web Service descriptions, querying the Semantic Web demands the added expressiveness provided by the instances. The ABox intelligent functions, such as instance checking, are of crucial importance when domain modelling calls for the fine-grained analysis that instances enable. This is especially true in an environment like the Web, where most of the semantic structures are of unspecified and arbitrary detail. In any case, inserting instances in an ontology enables inferences and expressions that would have been impossible to accomplish using concept classification solely.

The KDI presented in the following section utilizes therefore an *instance-based* method for conducting inferences that employs in its core two basic ABox functions: *instance checking*, that finds the concepts an instance belongs to and *role fillers retrieval*, which, given a specific instance and role, infers all related instances through this role. By utilizing these two features, in combination with the TBox functions, we can achieve not only the retrieval of information that has been explicitly expressed in the ontology, but also the discovery of knowledge that is logically deduced by this information. Therefore TBox as well as ABox support is an important criterion for selecting an appropriate reasoning back-end.

3.3 Choosing a Reasoning Back-End

Having chosen to use the DLs as the underlying formalism for our methodology, and having noted the most important characteristics a suitable implementation should have, we will now examine three inference engines based on DLs. Our evaluation is carried out in terms of expressiveness, support for OWL, reasoning about ABox and other main features provided by each of the three systems.

Cerebra, by Network Inference is a recent commercial system, providing reasoning as well as ontology management features. Cerebra supports nearly all constructors and axioms that would normally classify it to OWL DL expressiveness level. However, our experimentation with the system has shown some deficiencies which, in combination with its inability to reason about the ABox finally ranks Cerebra's expressiveness at $SHIQ$ level, at most. On the other hand, the relation model used by Cerebra allows the submission of very powerful queries, based on XQuery syntax. Still, the results are based only on the explicitly expressed information of the ontology, and not on information that may be inferred.

FaCT/FaCT++ [13] is a freely available reasoning software, that is being developed at Manchester. FaCT implements optimized, sound and complete algorithms to compute subsumption in the $SHIQ$(D) DL. FaCT does not support reasoning in ABox, neither concrete domains. It is also syntactically incompatible with OWL, since its knowledge bases are expressed in a Lisp-like or XML syntax.

FaCT++ [21] features greater expressiveness, aiming ultimately at OWL DL, by fully supporting concrete domains, while the underlying logic is $SHIF$(D). OWL syntax is not supported; however a transformation tool to the Lisp intermediate form is available. Individuals (and thus nominals) survive this transformation, but they are not yet fully supported, as they are approximated as primitive concepts.

RACER [10, 11] is an inference engine for very expressive DLs. It is the first system in its category to support reasoning in ABox as well as TBox, and this is its main asset in comparison to the other inference engines. RACER provides reasoning for $SHIQ(D)$, including instances (ABox). There is also full support for the OWL syntax. On the other hand, nominals are only approximated by creating a new concept for each of them.

RACER seems to be closer to the expressiveness needed by the Semantic Web mostly because of its enhanced support for OWL and its clear ability to reason about the ABox. Its utilisation in the KDI produced a number of interesting results, some of which are presented in Sect. 4.2.

4 The Knowledge Discovery Interface

In this section we describe the prototype Knowledge Discovery Interface. The KDI has been developed as an implementation of the decisions and criteria identified in the previous section. First we give a general description of the KDI along with a brief description of its functionality and the relation of its programmatic components to RACER functions. Then, we present two experimental inferences on OWL documents performed using the KDI and their results.

4.1 General Description and Architecture

The KDI is a web application, providing intelligent query submission services on Web ontology documents. We use the word *Interface* in order to emphasize the fact that the user is offered a simple and intuitive way to compose and submit queries. In addition, the KDI interacts with RACER to conduct inferences.

KDI is capable of loading ontologies of any structure and content. Furthermore, the user is capable of browsing the existing classes, their instances and their corresponding roles, thus presenting the user with all the needed information to compose his query. We have identified such a *declarative* behaviour to be of crucial importance for the Semantic Web knowledge discovery process; after all, the user should be able to pose queries even to unknown ontologies, encountered for the first time (Fig. 1).

KDI helps the user compose a query by selecting a concept, an instance and a role in a user friendly manner. After the query is composed, it is decomposed into several lower level functions that are then submitted to RACER. This procedure is transparent to the user, withholding the details of the knowledge base actual querying. This kind of *intuitive* composition of intelligent queries,

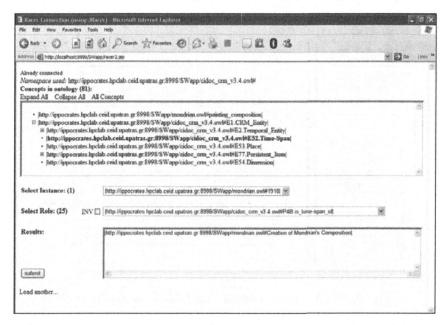

Fig. 1. Ontology classification and intelligent query composition using the KDI. The concept "E52.Time-Span" has been selected, thus only its instances and relevant roles are available. System correctly infers that "1918" is the year of the "Creation of Mondrian's composition"

significantly improves the knowledge discovery process by facilitating the user to pose precise and correct queries.

Finally, Fig. 2 displays the programming architecture of the KDI. This figure shows the various parts of the implemented business-logic (right part) as well as their relation to and dependence on RACER's lower level functions (left part): a functions pattern depicts the components it participates in.

4.2 Results

In the following we present the results from two different inference actions performed using the KDI, so as to demonstrate its capabilities as well as its limitations. In order to conduct these inferences we use the CIDOC Conceptual Reference Model [3, 4] as our knowledge base.

Firstly, we ported version 3.4 of the CRM to OWL format. Secondly we semantically enriched and extended CRM with concrete instances and more expressive structures, available only in OWL (like cardinality restrictions, inverse roles, existential and universal quantifications and so on). We then created a document named mondrian.owl that includes CRM concept and role instances which model facts from the life and work of the Dutch painter Piet Mondrian. In this document we also included axiom and fact declarations that

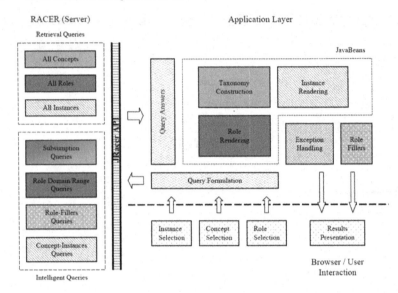

Fig. 2. The Architecture of the Knowledge Discovery Interface

OWL allows to be expressed, as well as new roles and concepts making use of this expressiveness.

The following code is a fragment from mondrian.owl stating that a "Painting_Event" is in fact a "Creation_Event" that "has_created" "Painting" objects only:

```
<owl:Class rdf:ID="Painting_Event">
<rdfs:subClassOf rdf:resource="&crm;E65.Creation_Event"/>
<rdfs:subClassOf>
<owl:Restriction>
<owl:onProperty rdf:resource="&crm;P94F.has_created"/>
<owl:allValuesFrom rdf:resource="#Painting"/>
</owl:Restriction>
</rdfs:subClassOf>
</owl:Class>
<Painting_Event rdf:ID="Creation of Mondrian's composition">
<crm:P94F.has_created rdf:resource="#Mondrian's composition"/>
</Painting_Event>
```

The above fragment is graphically depicted in the left part of Fig. 3. "Creation of Mondrian's Composition" (i_1) is an explicitly stated "Painting_Event" that "has_created" (R) "Mondrian's composition" (i_2). Now, asking the KDI to infer "what is a painting?" it infers that i_2 is indeed a painting (right part of Fig. 3), correctly interpreting the value restriction on role R.

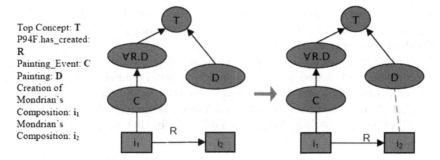

Fig. 3. Inference example using Value Restriction

Let's now examine another example that involves the use of nominals. The following fragment from mondrian.owl states that a "Painting" is a "Visual_Item" that its "Type" is "painting_composition".

```
<owl:Class rdf:ID="Painting">
<owl:subClassOf rdf:resource="&crm;E36.Visual_Item"/>
<owl:equivalentClass>
<owl:Restriction>
<owl:onProperty rdf:resource="&crm;P2F.has_type"/>
<owl:hasValue
rdf:resource="#painting_composition"/>
</owl:Restriction>
</owl:equivalentClass>
</owl:Class>
<crm:E55.Type rdf:ID="painting_composition"/>
<Painting rdf:ID="Mondrian's composition"/>
```

The above fragment is graphically depicted in the left part of Fig. 4.

"Mondrian's Composition" (i_1) is explicitly declared as a "Painting" instance which in turn is defined as a hasValue restriction on "has_type" (R). "Painting_composition" (i_2) is declared as a "Type" object. While the fact

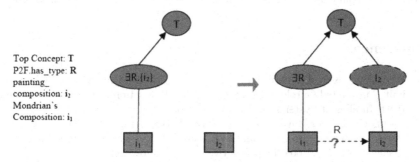

Fig. 4. Inference example using Existential Quantification and nominals

that "Mondrian's Composition" "has_type" "Painting" is straightforward, the KDI is unable to infer so and returns *null* when asked "what is the type of Mondrian's composition?"

This example clearly demonstrates the inability of RACER as well as every other current DL based system to reason about nominals. Given the $\{i_2\}$ nominal, RACER creates a new synonym concept I_2 and makes i_2 an instance of I_2. It then actually replaces the hasValue restriction with an existential quantifier on *concept I_2* and thus is unable to infer that $R(i_1, i_2)$ really holds.

5 Conclusions

In this paper we have shown how the Semantic Web may fulfill a large part of what it promises, mainly through its knowledge discovery features. These features are grounded on a well-studied background and their adaption to the Web environment is now mature and even standardized to some extent. The KDI and its underlying methodology demonstrate proper evidence of how these features can be practically applied so as to be beneficial for a number of applications.

Our proposed methodology comes to fill the gap between existing approaches by designating a process that can be followed to achieve knowledge discovery on the Semantic Web. Our choices depend on the current state-of-the-art in inference engines and Semantic Web standardization efforts. We trust that at the near future most of the difficulties and incompatibilities identified throughout our work would be overridden by the evolution of systems and the refinement and possibly enrichment of the Ontology Web Language.

The KDI for example is greatly hampered by the limited expressiveness and scalability of current DL inference engines, regarding the use of nominals and the processing of large ontology documents respectively. Despite that fact, the KDI displays a combination of innovations and distinctive features that are not, to our knowledge, simultaneously met by any other system. Among the most important of them is an intuitive and declarative way of constructing and submitting queries and the implementation of an original inference methodology that is especially suitable for the Semantic Web environment.

References

1. H. Alani, S. Kim, D.E. Millard, M.J. Weal, W. Hall, P.H. Lewis and N.R. Shadbolt. Automated Ontology-Based Knowledge Extraction from Web Documents. IEEE Intelligent Systems, 18(1): 14–21, 2003.
2. Bechhofer, F. van Harmelen, J. Hendler, I. Horrocks, D.L. McGuinness, P.F. Patel-Schneider and L.A. Stein. OWL Web Ontology Language Reference. W3C Recommendation, 2004. http://www.w3.org/TR/owl-ref/

3. N. Crofts, M. Doerr and T. Gill. The CIDOC Conceptual Reference Model: A standard for communicating cultural contents. Cultivate Interactive, issue 9, 2003. http://www.cultivate-int.org/issue9/chios/

4. M. Doerr. The CIDOC conceptual reference module: an ontological approach to semantic interoperability of metadata. AI Magazine, 24(3): 75–92, 2003.

5. R. Fikes, P. Hayes and I. Horrocks. DQL – A Query Language for the Semantic Web. KSL Technical Report 02–05, 2002.

6. R. Fikes, P. Hayes and I. Horrocks. OWL-QL: A Language for Deductive Query Answering on the Semantic Web. KSL Technical Report 03–14, 2003.

7. R. Fikes, J. Jenkins, and F. Gleb. JTP: A System Architecture and Component Library for Hybrid Reasoning. In Proc. of the Seventh World Multiconference on Systemics, Cybernetics, and Informatics, 2003.

8. J. González-Castillo, D. Trastour and C. Bartolini. Description Logics for Matchmaking of Services, In Proc. of KI-2001 Workshop on Applications of Description Logics, 2001.

9. R. Guha, R. McCool. TAP: A Semantic Web Platform. Computer Networks, 42(5):557–577, 2003.

10. V. Haarslev and R. Möller. Racer: A Core Inference Engine for the Semantic Web. In Proc. of the 2nd International Workshop on Evaluation of Ontology-based Tools (EON2003), pp. 27–36, 2003.

11. V. Haarslev and R. Möller. RACER User's Guide and Reference Manual Version 1.7.19. http://www.sts.tu-harburg.de/~r.f.moeller/racer/racer-manual-1-7-19.pdf

12. I. Horrocks and P.F. Patel-Schneider. Reducing OWL entailment to description logic satisfiability. In D. Fensel, K. Sycara, and J. Mylopoulos (eds.): Proc. of the 2003 International Semantic Web Conference (ISWC 2003), number 2870 of LNCS, pp. 17–29. Springer, 2003.

13. I. Horrocks and U. Sattler. Optimised reasoning for SHIQ. In Proc. of the 15th Eur. Conf. on Artificial Intelligence (ECAI 2002), pp. 277–281, 2002.

14. I. Horrocks and S. Tessaris. Querying the Semantic Web: a Formal Approach. In I. Horrocks and J. Hendler (eds.): Proc. of the 13th Int. Semantic Web Conf. (ISWC 2002), number 2342 in LNCS, pp. 177–191. Springer, 2002.

15. E. Hsu and D. McGuinness. Wine Agent: Semantic Web Testbed Application. In Proc. Of Workshop on Description Logics, 2003.

16. J. Kopena and W.C. Regli. DAMLJessKB: A tool for reasoning with the Semantic Web. IEEE Intelligent Systems, 18(3): 74–77, 2003.

17. L. Li and I. Horrocks. A software framework for matchmaking based on semantic web technology. In Proc. of the Twelfth International World Wide Web Conference (WWW 2003), pp. 331–339. ACM, 2003.

18. J. Mayfield and T. Finin. Information retrieval on the Semantic Web: Integrating inference and retrieval. In Proc. of SIGIR Workshop on the Semantic Web, 2003.

19. Network Inference Ltd. Description Logics (white paper), http://www.networkinference.com

20. M. Paolucci, T. Kawamura, T.R. Payne and K. Sycara. Semantic Matching of WebServices Capabilities. In Proc. of International Semantic Web Conference (ISWC), 2002.

21. D. Tsarkov and I. Horrocks, Reasoner Prototype: Implementing new reasoner with datatype support. IST-2001-33052 WonderWeb Del. 13, http://wonderweb.semanticweb.org